T0288949

ESTIMATION AND PREDICTION OF BALLISTIC MISSILE TRAJECTORIES

Jeffrey A. Isaacson
David R. Vaughan

Prepared for the United States Air Force
Approved for public release; distribution unlimited

RAND

The research described in this report was sponsored by the United States Air Force under Contract F49620-91-C-0003. Further information may be obtained from the Strategic Planning Division, Directorate of Plans, Hq USAF.

ISBN: 0-8330-2376-4

RAND is a nonprofit institution that helps improve public policy through research and analysis. RAND's publications do not necessarily reflect the opinions or policies of its research sponsors.

Published 1996 by RAND
1700 Main Street, P.O. Box 2138, Santa Monica, CA 90407-2138
RAND URL: http://www.rand.org/
To order RAND documents or to obtain additional information,
contact Distribution Services: Telephone: (310) 451-7002;
Fax: (310) 451-6915; Internet: order@rand.org

This report documents analysis originating from more comprehensive RAND research to establish an investment strategy for U.S. space systems and concepts of operation for countering critical mobile targets. The work was conducted within the Project AIR FORCE Force Modernization and Employment program, under the auspices of the C^4I/Space project, for the Air Combat Command.

The study describes an analytical tool useful in establishing figures of merit for satellites in a notional operational setting in which ballistic missile defenses are employed. A framework familiar to system designers is described pedagogically, and its utility in deriving operational implications is demonstrated for one interesting case. The report should be useful to decisionmakers and analysts within the U.S. Air Force and the Department of Defense, as well as others generally concerned with theater missile defense architectures and operational effectiveness analysis.

PROJECT AIR FORCE

Project AIR FORCE, a division of RAND, is the Air Force federally funded research and development center (FFRDC) for studies and analyses. It provides the Air Force with independent analyses of policy alternatives affecting the development, employment, combat readiness, and support of current and future aerospace forces. Research is performed in three programs: Strategy, Doctrine, and Force Structure; Force Modernization and Employment; and Resource Management and System Acquisition.

Project AIR FORCE is celebrating 50 years of service to the United States Air Force in 1996. Project AIR FORCE began in March 1946 as Project RAND at Douglas Aircraft Company, under contract to the Army Air Forces. Two years later, the project's contract and personnel were separated from Douglas to form a new, private nonprofit institution to improve public policy through research and analysis for the public welfare and security of the United States—the foundation of what is known today as RAND.

CONTENTS

FIGURES

TABLES

Thirty-three nations, a number of which actively pursue policies contrary to U.S. interests, possess TBMs. Moreover, the exportable supply of TBMs continues to grow through worldwide development efforts, and missiles of increased range and payload could find their way into the weapons inventories of many nations during the next decade. Coupled with a concomitant spread of weapons of mass destruction (WMD), such TBMs could permit a strike capability that could threaten regional balances, U.S. allies, or even U.S. forces deployed overseas. Thus, although there are diplomatic efforts to curtail missile proliferation,[1] the United States has undertaken an ambitious research and development effort in theater missile defense (TMD).

Active defenses, passive defenses, attack operations, and command, control, communications, and intelligence (C3I) form the four "pillars" of the U.S. theater defense program.[2] As theater missile defenses are fielded at the decade's end, satellite sensors will likely play an important supporting role. How might these sensors contribute to C3I in the TMD environment?

---

[1]The Missile Technology Control Regime (MTCR) is one such effort. Created in 1987, the MTCR controls the transfer of technologies that could aid the unmanned delivery of a 500-kilogram payload over a 300-kilometer distance. For a brief description of the MTCR, see Ballistic Missile Defense Organization, *Ballistic Missile Proliferation: An Emerging Threat*, Arlington, Virginia: System Planning Corporation, 1992, pp. 64–65.

[2]C3I is in a sense the foundation supporting these pillars, rather than a pillar itself.

Consider the notional missile launch depicted in Figure S.1. A satellite sensor in position to view a boosting TBM[3] can in principle provide useful information to a variety of theater defense platforms. By gathering information on the TBM trajectory, for example, a "forward track" of the missile can be derived, enabling the time and location of missile impact to be estimated. If relayed to the target area in a timely manner, appropriate passive defensive measures may be employed. In addition, the forward track can include estimates of the missile position as a function of time along the trajectory. Such

RAND*MR737-S.1*

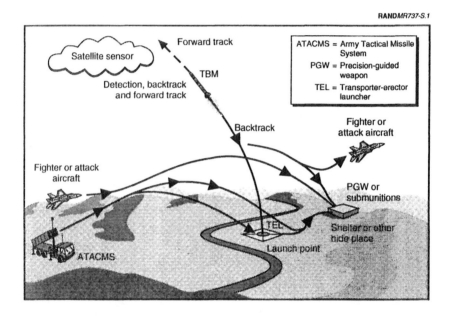

Figure S.1—Satellite Sensors Support Both Forward and Target Area Defenses

---

[3]To simplify our discussion, we use the term "satellite sensor" to represent a spaceborne platform capable of detecting missiles during the boost-phase only. Sensors capable of detecting TBMs after booster burnout (e.g., Brilliant Eyes-type systems) are not considered here.

estimates could be used to cue search radars of active defense systems, and perhaps provide fire-control quality "launch baskets" for TBM interceptors.[4]

Similarly, a TBM "backtrack" to the launch point provided by satellite sensors could support attack operations with aircraft or ground-launched weapons. During the Gulf War, Scud launchers could be moved within minutes of missile firing, and after 15 minutes could be anywhere within nine miles of the launch point, underscoring the importance of timely response.[5] By detecting and tracking the TBM during boost-phase,[6] however, the spaceborne systems considered here have the potential to supply information for such a response, and to do so nearly globally on an essentially continuous coverage basis.

To examine the capabilities satellites bring to bear in the TMD environment, we describe a filtering methodology for the estimation and prediction of ballistic missile trajectories[7] and apply it to a notional TBM with a boost-phase of 100-sec duration and a total range of 1200 km. During the estimation sequence, measurements of the missile trajectory are obtained from an assumed template,[8] constructed by modeling the missile's flight in the atmosphere of a spherical, non-rotating earth. The state vector we estimate is defined in six dimensions, with elements representing the missile's latitude, longitude, heading, time, and altitude at launch, as well as its loft angle during boost-phase. We assume a launch in Iran (at 34.01° latitude, 47.40° longitude) with a 263° heading,[9] impacting Tel Aviv at 32.05° lati-

---

[4]In the case of boost-phase/ascent-phase intercept, time constraints may limit the utility of satellite-based information.

[5]Secretary of Defense, *Conduct of the Persian Gulf War: Final Report to Congress*, Washington, D.C.: U.S. Government Printing Office, April 1992, p. 224.

[6]Depending on the type of missile, boost-phases typically last between 30 and 120 sec. See Congressional Budget Office, *The Future of Theater Missile Defense*, Washington, D.C.: U.S. Government Printing Office, June 1994, p. 5.

[7]While particulars may vary, a similar methodology is likely to be used in any operational system that is tasked with TBM trajectory analysis.

[8]We simulate the measurement process in order to estimate the errors one might expect using the filter technique. In the field, measurements would be obtained from the actual missile under observation. See Chapter Three for a more detailed description.

[9]0° represents due north and 90° represents due east.

tude, 34.77° longitude. The relevant geometry is illustrated in Figure S.2, where the satellites are positioned in geosynchronous orbit at 0° latitude, and 15° and 75° east longitude, respectively.[10]

The two satellites sample the trajectory independently, each measuring two angles (which are subject to random errors and bias) at 20-sec intervals (the assumed revisit time[11]). To begin filtering, we must specify the initial covariance of the state estimate error before measurement.[12] We assume an initial 1° uncertainty in launch latitude and longitude, 20° uncertainty in launch heading, 20-sec uncertainty in launch time, 1-km uncertainty in launch altitude, and a 1°

RANDMR737-S.2

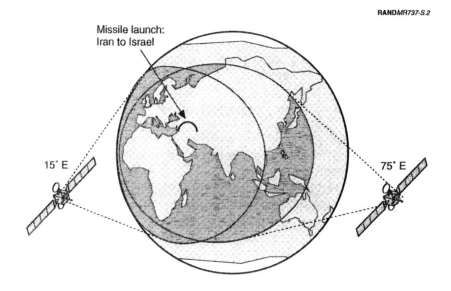

Figure S.2—Geometry of TBM Trajectory and Sensors

---

[10]Although we illustrate the methodology for a notional missile launch and satellite configuration, the formulas and equations derived are generally applicable to a wide range of threat scenarios and sensor constructs.

[11]In a sensitivity excursion, we later examine the effects on the trajectory analysis of varying the revisit time.

[12]See Chapter Two for a detailed description of the estimation sequence.

uncertainty in loft angle. For simplicity, we assume clouds do not present a viewing problem, and ignore effects of early booster engine cutoff.[13]

NOTIONAL RESULTS: RANDOM ERRORS ONLY

One useful application of the filtering methodology is in estimating the uncertainty associated with the location of a TBM launch. As depicted in Figure S.3, launch point uncertainty (LPU) may be reduced significantly by processing the measurements from both satellites sequentially (i.e., stereo processing). (Here, $\ell = 2$ corresponds to an 87 percent confidence level.[14]) The LPU derived monoscopically from each separate sensor is also shown, indicating how a different viewing geometry may lead to different results.

In the absence of measurement errors, six measured angles would uniquely determine the six-dimensional state vector we are estimating (assuming our template is exact). Since each measurement provides two angles, only three measurements would be required to determine the state.

Figure S.4 illustrates the $\ell = 2$ launch point uncertainty as a function of time for various random errors in measurement angle (100, 30, and 10 microradians) and stereo processing. As is clear in all cases, a priori uncertainties are reduced most rapidly by the first few measurements, and at a slower pace thereafter. (Measurements occur at times indicated by dots in the figure.) As expected on intuitive grounds, moreover, the LPU derived after the final measurement has been made scales roughly as the square of the random error.

Determining the uncertainty associated with missile location at any point along its trajectory is another useful application of the technique. Missile location uncertainties (MLUs)[15] for two sensors pro-

---

[13]That is, we assume full-burn trajectories throughout. Note, however, that the time at burnout is still uncertain, owing to missile launch time uncertainty.

[14]A discussion of probabilities and uncertainty ellipses is found in Chapter Three.

[15]MLU is the volume of an ellipsoid that surrounds the estimated target position and contains the actual target with some specified probability.

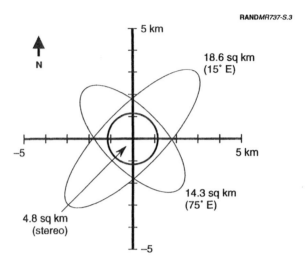

Figure S.3—LPUs (ℓ = 2) for Two Sensors with Random Errors

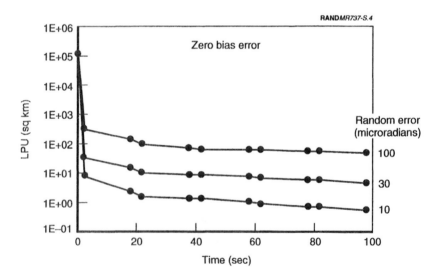

Figure S.4—Sensitivity of LPU (ℓ = 2) to Random Error (Two Sensors)

cessed stereoscopically are shown in Figure S.5 for various random errors. (Here, in three dimensions $\ell = 2$ corresponds to a 74 percent confidence level.) As illustrated, the uncertainty volume increases monotonically until the latter part of the trajectory, when the MLU turns over.[16] (As a point of reference, a sphere of 62-km radius encloses a volume of roughly 10^6 km^3.)

In general, decreasing the revisit time allows more measurements to be made and, consequently, provides more information about the missile trajectory. Figure S.6 illustrates the LPU for various revisit times, spanning the range of 2.5–40 sec. At late times, note that the LPU scales roughly linearly with the number of measurements.

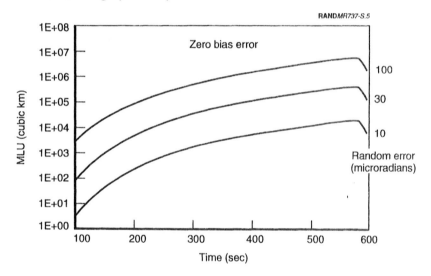

Figure S.5—Sensitivity of MLU (ℓ = 2) to Random Error (Two Sensors)

---

[16]By examining the trajectories with perturbed launch times, altitude, and loft, one finds for the example at hand that the deviation from the nominal baseline trajectory begins to decrease at an altitude of approximately 30,000 ft. This effect, manifested in the decreasing uncertainty 580 sec into the flight, is related to both the atmospheric degradation of the missile velocity upon reentry and our choice of a minimum energy trajectory to perturb about.

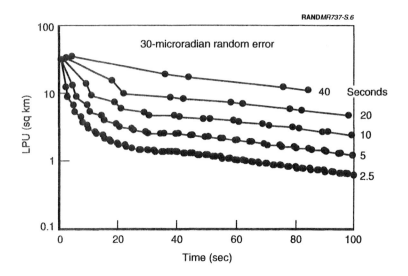

Figure S.6—Sensitivity of LPU (ℓ = 2) to Revisit Time (Two Sensors)

Figure S.7 shows the effect of revisit time on missile location uncertainty. As is evident from this plot, an order-of-magnitude reduction in revisit time generates more than an order-of-magnitude reduction in uncertainty volume.

NOTIONAL RESULTS: RANDOM AND BIAS ERRORS

At this point, we have considered a filter optimized for random errors alone. In many situations, however, bias errors dominate the measurement uncertainty, and must therefore be accounted for. There are two possibilities: (1) examine the effect of bias errors on the existing filter optimized for random errors, and (2) design a filter to account for the bias errors explicitly. We refer to these formulations as suboptimal and optimal, respectively.[17]

---

[17]It is natural to ask why one would bother using a suboptimal formulation. If redesigning an existing filter optimized for random errors alone is not desirable, the suboptimal approach allows the effect of bias on that filter to be examined, albeit as an afterthought. See Chapter Four for a detailed description of both approaches.

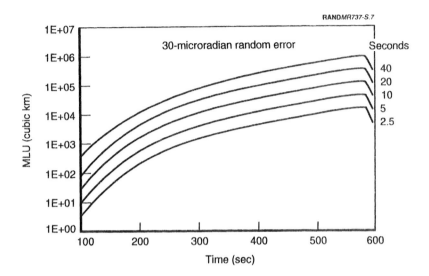

Figure S.7—Sensitivity of MLU (ℓ = 2) to Revisit Time (Two Sensors)

When treating bias suboptimally, the filter applies gains—indeed, sometimes large gains—to the system by considering random errors alone. As a result, when the effects of bias are examined, they may be large because they are amplified by large gains. In an optimal formulation, on the other hand, the filter knows bias errors are present and can adjust these gains accordingly. Nonetheless, when the bias is not dominant (i.e., bias error is less than or comparable to the random error), one would expect both approaches to yield similar results.[18]

For the notional TBM launch described above, Figure S.8 illustrates the launch point uncertainty in the presence of bias (treated optimally). Relative to Figure S.3, larger launch point uncertainty ellipses are obtained with both sources of error present.

---

[18]Keep in mind, however, that the suboptimal treatment of bias can go awry even in cases where the random and bias errors are comparable. If the filter applies large gains during the estimation sequence, results obtained treating bias suboptimally may differ markedly from those obtained with an optimal formulation. In all cases, though, the former approach would *overestimate* the error.

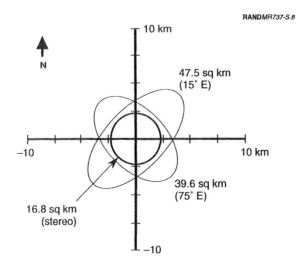

RANDMR737-S.8

30-microradian random and bias errors

Figure S.8—LPUs (ℓ = 2) for Two Sensors with Random and Bias Errors (Optimal Filter)

Figure S.9 illustrates the time evolution of the launch point uncertainty for the case of a 30-microradian random error and 100-, 30-, and 10-microradian bias errors, respectively. Unlike the case with random errors alone (Figure S.4), the LPU derived after the final measurement does not scale as the square of the error. Moreover, the importance of bias is apparent in the large difference between the 30- and 100-microradian cases.

The missile location uncertainty, illustrated in Figure S.10, also shows qualitative differences from estimates constructed in the absence of bias (Figure S.5). In particular, curves characterized by different biases appear to coalesce late in the trajectory.

Finally, as sensor revisit times are varied, the LPU derived after the final measurement is relatively insensitive to revisit time in the case of 30-microradian random and bias errors (Figure S.11). In effect, a

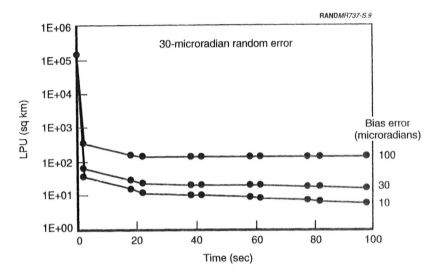

Figure S.9—Sensitivity of LPU (ℓ = 2) to Bias Error (Two Sensors; Optimal Filter)

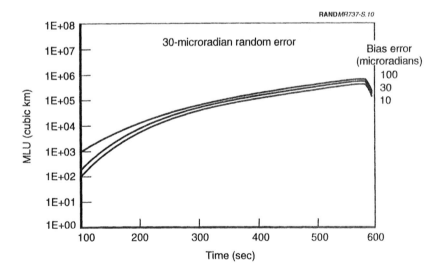

Figure S.10—Sensitivity of MLU (ℓ = 2) to Bias Error (Two Sensors; Optimal Filter)

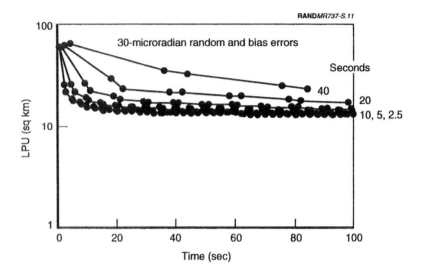

Figure S.11—Sensitivity of LPU ($\ell = 2$) to Revisit Time (Two Sensors; Optimal Filter)

point is reached in the filtering sequence where additional measurements containing unknown bias errors provide information of limited utility. This underscores the point that *increasing the data collection rate will not reduce the launch point uncertainty significantly unless random errors dominate the measurement process.* This is because statistics alone do not "beat down" the effects of bias. Although the insensitivity to revisit time is not apparent in the case of missile location uncertainty (Figure S.12), the spread in MLU values as revisit time is varied is reduced relative to the unbiased case (Figure S.7).

On the other hand, this is not to suggest that revisit time is wholly unimportant in the presence of bias errors. In the event of early booster engine cutoff, for example, sizable uncertainties in burnout velocity could dominate the error analysis—with or without bias effects. By using the general method described herein—which can ac-

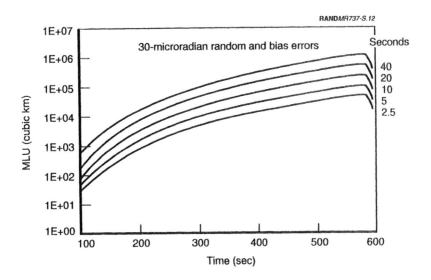

RAND*MR737-S.12*

Figure S.12—Sensitivity of MLU ($\ell = 2$) to Revisit Time (Two Sensors; Optimal Filter)

commodate, and thus estimate, the effects of early booster engine cutoff—a short revisit time could improve our knowledge of the missile burn time, and, consequently, our state vector estimate.

CONCLUDING REMARKS

As theater missile defenses are fielded at the decade's end, satellite sensors will likely realize an important TMD battle management function. Waging "information warfare" will require increasingly so-phisticated C^3I networks that can piece together the multifarious packets of information required to effect battlespace dominance. In this regard, timely transmission throughout the theater is central. But successful battle management requires more than connectivity alone: the *quality* of the information being transmitted is para-mount. Thus, our primary focus in this study has been on describing a technique whose application can in principle provide such information in the TMD operational environment.

Harnessed in a theater of operations, the type of information described here can be used to enhance the capability of active defenses, passive defenses, and attack operations. It is thus important for the Air Force to model and understand such enhancement in operational terms, so that personnel can understand the trade-offs available between revisit time and high accuracy. Indeed, the use of models that can capture the operational effects of these technical details seems important for any decisions involving the acquisition of space-based sensor systems. As the data presented here demonstrate, this is especially true of sensors with short revisit times and small measurement errors, at least insofar as our notional trajectory analysis is concerned. It is important to remember, however, that bias errors can be significant, and perhaps even dominant. Including their effects (optimally in certain circumstances) is therefore central to the success of any methodology seeking to estimate and predict ballistic missile trajectories. Moreover, techniques to reduce or eliminate such errors, where applicable, should be given due consideration.

ACKNOWLEDGMENTS

The authors wish to thank Michael Jacobs and Howard Holtz of the Aerospace Corporation for sharing their insights on trajectory estimation and prediction; RAND colleagues James Bonomo and Michael D. Miller for thoughtful reviews; Herbert Hoover, Mario Juncosa, and Moira Regelson for helpful suggestions; Richard Buenneke for graphical assistance; June Kobashigawa for manuscript preparation; and finally, Stephen Guarini for invaluable Fortran programming support at the project's outset. Needless to say, responsibility for any errors or omissions is our own.

ACRONYMS

ATACMS	Army Tactical Missile System
BMD	Ballistic missile defense
C^3I	Command, control, communications, and intelligence
CONOPS	Concepts of operation
DoD	Department of Defense
ERINT	Extended-range interceptor
JSTARS	Joint Surveillance Target Attack Radar System
km	Kilometer
LPU	Launch point uncertainty
µrad	Microradian
MEADS	Medium Extended Air Defense System
MLU	Missile location uncertainty
MTCR	Missile Technology Control Regime
ODS	Operation Desert Storm
PAC	Patriot Advanced Capability
PGW	Precision-guided weapon
RV	Reentry vehicle
rpm	Rotations per minute
SAM	Surface-to-air missile
SDI	Strategic Defense Initiative
TBM	Theater ballistic missile
TBMD	Theater ballistic missile defense
TEL	Transporter-erector-launcher
THAAD	Theater High-Altitude Area Defense
TMD	Theater missile defense
TMD-GBR	Theater missile defense ground-based radar
WMD	Weapons of mass destruction

Chapter One

INTRODUCTION

At the outset of Operation Desert Storm (ODS), actively defending against ballistic missile attack was not a new idea. Indeed, the U.S. Air Force had begun examining the technical feasibility of ballistic missile defense (BMD) as early as 1946 with projects Wizard and Thumper, before many relevant technologies were mature enough to offer much hope for success. Recognizing the similarity between air defense and missile defense, the U.S. Army entered the BMD arena in 1955, when it began developing Nike-Zeus, a nuclear-tipped interceptor based on the Nike-Hercules anti-aircraft system. By 1958, an interservice competition for the BMD mission was well under way.[1]

At the same time, new technical issues arose that called the BMD mission into question: Could radars discriminate between reentry vehicles (RVs) and decoys above the atmosphere? Would the system become saturated if RVs arrived at close intervals? Was guidance adequate to bring the interceptor to within the kill radius? Could the system function properly in a nuclear environment?[2] Largely be-

---

[1]D. N. Schwartz, *Past and Present: The Historical Legacy*, in A. Carter and D. N. Schwartz (eds.), *Ballistic Missile Defense*, Washington, D.C.: The Brookings Institution, 1984, pp. 331–332.

[2]We note that the contextual setting of early BMD work was much different than that of today, and thus research efforts faced different problems. For example, the nuclear threat mandated low leakage levels and required systems to be functional in a nuclear environment. Moreover, strategic competition with the Soviet Union often made technical issues (e.g., decoy discrimination) hard to settle. While many of these issues persist, the present context renders their resolution less crucial.

cause of these concerns, Nike-Zeus production stagnated throughout the Eisenhower years.[3]

By 1963, technological advances in the areas of computing, radar, and propulsion established the feasibility of an endoatmospheric interceptor, Nike-X, which could in principle discriminate between RVs and decoys by discerning differences in their interactions with the atmosphere. With phased-array radars, moreover, the system would be less vulnerable to saturation. Despite these advantages, Nike-X (later called Sentinel) became vulnerable to a new set of strategic considerations first raised by the McNamara Pentagon: the prospect that missile defenses could stimulate a destabilizing arms race with the Soviet Union. Thus, Sentinel was suspended in 1969 by the Nixon administration, and although its revised BMD program (Safeguard) was initially funded, by May 1972 the United States and the Soviet Union had established a treaty aimed at limiting the development of ballistic missile defenses to very low levels. By congressional directive, Safeguard was terminated in fiscal 1976.[4]

On March 23, 1983, a speech by President Ronald Reagan brought BMD to the fore of public consciousness and set in motion an extensive research and development effort known as the Strategic Defense Initiative (SDI). Harnessing new technological achievements, SDI sought to provide a defensive umbrella shielding the United States from strategic attack. In the ensuing years, the conceptual and technical feasibility of BMD was revisited in a new context, although many issues remained unresolved.[5] But because of cost, the warming of superpower relations in the late 1980s, and long-standing concerns that BMD could undermine a relatively stable strategic balance, the initial fervor associated with SDI waned by the decade's end. Nevertheless, thinking about missile defense was alive in 1990, albeit focused primarily on protecting the U.S. homeland.[6]

---

[3]Ibid., pp. 332–333.

[4]Ibid., pp. 334–344.

[5]A broad collection of essays on this subject is found in A. Carter and D. N. Schwartz, 1984.

[6]Defending against conventionally armed Soviet missiles in Europe was one exception. Indeed, had this not been an issue in the mid-1980s, the Patriot missile deployed in ODS may not have had any capability to engage TBMs.

Refocusing this thinking on protecting U.S. forces and allies in operational theaters arguably began on the second day of ODS, when modified Scud missiles landed on Tel Aviv. Although few people were injured in the initial attacks, the spectre of chemical weapons threatened to draw Israel into the Gulf conflict, potentially undermining a somewhat fragile coalition of Arab states allied with the United States against Iraq. It became apparent, consequently, that theater ballistic missile (TBM) use could exact a heavy toll in the political arena, if not the operational one.

As the war continued on, however, the toll of TBM strikes on tactical operations also became apparent. "Scud-hunting" with F-15Es, F-16s, A-10s, A-6Es, B-52s, and JSTARS aircraft diverted thousands of air sorties away from other missions. Reconnaissance aircraft (U-2/TR-1s and RF-4Cs) were also shifted.[7] Although the defensive performance of Patriot missiles provided a positive psychological factor, it became clouded in controversy[8] and contributed to the substantial property damage inflicted by the 88 modified Scuds launched during the war.[9] Finally, 28 U.S. soldiers were killed in Dhahran, Saudi Arabia, when a single TBM struck their barracks.

In large measure, the ODS experience galvanized U.S. interest in theater missile defense (TMD), in part because of the world's sizable inventory of ballistic missiles. Thirty-three nations, a number of which actively pursue policies contrary to U.S. interests, possess TBMs. (See Figure 1.1.)[10]

---

[7]Secretary of Defense, *Conduct of the Persian Gulf War: Final Report to Congress*, Washington, D.C.: U.S. Government Printing Office, April 1992, pp. 224–226.

[8]For example, see, T. A. Postol, "Lessons of the Gulf War Experience with Patriot," *International Security*, Vol. 16, No. 3, Winter 1991/92, pp. 119–171; R. M. Stein, "Patriot ATBM Experience in the Gulf War," *International Security*, Vol. 16, No. 3, Winter 1991/92, addendum; R. M. Stein and T. A. Postol, "Correspondence: Patriot Experience in the Gulf War," *International Security*, Vol. 17, No. 1, Summer 1992, pp. 199–240.

[9]See Secretary of Defense, 1992, pp. 226–227; S. Fetter, G. N. Lewis, and L. Gronlund, "Why Were Scud Casualties So Low?" *Nature*, 28 January 1993, pp. 293–296.

[10]These missiles are in service and have maximum ranges of 200 kilometers or greater. The "former USSR" in Figure 1.1 includes only Azerbaijan, Belarus, Georgia, Kazakhstan, Russia, and Ukraine. See D. Lennox, "Ballistic Missiles Hit New Heights," *Jane's Defence Weekly*, 30 April 1994, pp. 24–28. For a broader discussion of ballistic missile proliferation, see Janne E. Nolan, *Trappings of Power: Ballistic Missiles in the Third World*, Washington, D.C.: The Brookings Institution, 1991.

RAND*MR737-1.1*

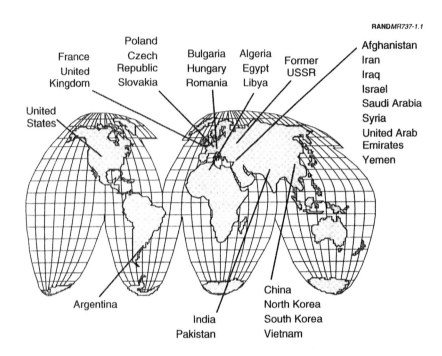

Figure 1.1—Thirty-Three Nations Possess TBMs

Perhaps more important, worldwide development efforts contribute to the exportable supply of TBMs, many of which may realize maximum ranges in excess of Iraq's 650-km[11] Al Hussein (see Table 1.1). Coupled with a concomitant spread of weapons of mass destruction (WMD), such TBMs could enable a strike capability that might threaten regional balances, U.S. allies, or even U.S. forces deployed overseas. The evolving security environment contains elements that are potentially worrisome at best; at worst, they are downright threatening.

----

[11]D. Lennox, 1994.

Table 1.1

Development Programs Complicate Efforts to Curtail TBM Proliferation

Country	Missile	Range (km)	Payload (kg)
Iran	Mushak 200	200	500
South Korea	NHK/A (Hyon Mu)	300	300
Pakistan	Hatf 2	300	500
Iran	CSS-7/M-11 variant	300	500
India	Prithvi SS-350	350	500
Pakistan	Hatf 3	600	500
Iran	Iran 700 (Scud C)	700	500
Libya, Iran	Al Fatah	950	500
Taiwan	Tien Ma (Sky Horse)	950	500
North Korea, Iran	Labour-1 (Nodong 1)	1000	1000
China, Iran	M 18 (Tondar-68)	1000	400
Spain	Capricornio	1300	500
North Korea, Iran	Labour-2 (Nodong 2)	1500	1000
China, Iran	DF-25	1700	2000
North Korea	Taepo-Dong 1	2000	1000
India	Agni	2500	1000
North Korea	Taepo-Dong 2	3500	1000

SOURCE: D. Lennox, "Ballistic Missiles Hit New Heights," *Jane's Defence Weekly*, 30 April 1994, pp. 24–28.

TMD DEVELOPMENT IS UNDER WAY

Notwithstanding diplomatic efforts to curtail missile proliferation,[12] it is no surprise that the United States has undertaken an ambitious research and development effort in theater missile defense. To better understand the TMD mission, a notional "cradle to grave" TBM deployment sequence is illustrated in Figure 1.2, along with the "Core Systems" planned by the Department of Defense (DoD) (discussed below).

---

[12]The Missile Technology Control Regime (MTCR) is one such effort. Created in 1987, the MTCR controls the transfer of technologies that could aid the unmanned delivery of a 500-kilogram payload over a 300-kilometer distance. For a brief description of the MTCR, see Ballistic Missile Defense Organization, *Ballistic Missile Proliferation: An Emerging Threat*, Arlington, Virginia: System Planning Corporation, 1992, pp. 64–65.

RAND*MR737-1.2*

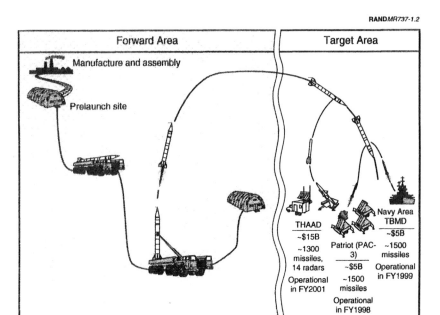

Figure 1.2—Core Systems Emphasize Target Area

Following its manufacture/assembly in a production facility, the notional missile is transported to a prelaunch site, which may be difficult to locate and destroy. When a deployment order is given, the TBM moves on a transporter-erector-launcher (TEL) to the launch site, where the missile is erected and fired. Following a period of powered flight in which rocket fuel burns with a bright signature, the missile proceeds on a ballistic trajectory, defined in large measure by its velocity and position at burnout. At this point, the missile is on its way to impact, and the mobile TEL may be fleeing to a postlaunch "hide site" or resupply depot.

It is convenient to differentiate between opportunities to counter TBMs in the target area and the forward area (Figure 1.2). Target area defenses either attempt to intercept the incoming missile[13] near

---

[13]As the situation dictates, reentry vehicles may be the appropriate targets, rather than the missiles themselves.

the point of impact (the Patriot missile in the Gulf War is a familiar example of such a system) or rely on passive measures in the target zone (seeking shelter, donning protective clothing, etc.). The TMD Core Systems currently planned by DoD emphasize interception for target area defense, with three separate initiatives: PAC-3 (with extended-range interceptor, ERINT),[14] Navy Area TBMD,[15] and Theater High-Altitude Area Defense (THAAD) (with TMD-GBR).[16] These systems are scheduled for initial deployment in 1998, 1999, and 2001, respectively, at a total cost of about $25 billion.[17]

Forward area defenses, on the other hand, would target the missile while it is boosting or ascending, thereby providing capability against TBMs with fractionating payloads.[18] Production facilities, prelaunch sites, resupply depots, and the TEL itself could also be targeted. Attack operations of this sort—and, indeed, forward area defenses in general—would likely employ aircraft, owing to the need to reach into the forward area.[19] Although forward area development

---

[14]Patriot Advanced Capability-3 (with extended-range interceptor) is, roughly speaking, a new and improved Patriot missile. Existing Patriot launchers and radars will be modified.

[15]Navy Area TBMD (formerly known as Navy Lower-Tier) will use Standard Block IVA missiles deployed on roughly 50 AEGIS cruisers and destroyers. Ship-based radars will be modified to accommodate the TMD mission.

[16]THAAD (with TMD ground-based radar) is a ground-based, upper-tier defense system requiring new missiles and new radars for target acquisition and fire control.

[17]See Congressional Budget Office, *The Future of Theater Missile Defense*, Washington, D.C.: U.S. Government Printing Office, June 1994, p. xv. The above cost includes estimates of funds appropriated before 1995.

[18]Boost-phase interceptors are one of three Advanced-Capability TMD Systems currently being examined by DoD—Navy Theater-Wide TBMD (formerly known as Navy Upper-Tier) and the Medium Extended Air Defense System [MEADS] (formerly known as Corps SAM) are the others. Because of budgetary constraints, it is expected that only one of these will eventually proceed to development. See Congressional Budget Office, 1994, p. xiv.

[19]Special Operations Forces (SOF) deployed in the forward area could support attack operations by relaying information about TBM launches to strike aircraft. During ODS, SOF groups in fact provided vital information about Iraqi missiles. See Secretary of Defense, 1992, p. 226; and D. C. Waller, *The Commandos*, New York: Simon & Schuster, 1994, pp. 335–351.

programs have not received the highest priority within DoD, promising concepts of operation (CONOPS) have been identified.[20]

SATELLITE SENSORS SUPPORT TMD BATTLE MANAGEMENT

Active defenses, passive defenses, and attack operations as described above form three of the four "pillars" of the U.S. theater defense program. The fourth—command, control, communications, and intelligence (C^3I)—is in a sense the foundation supporting these pillars, rather than a pillar itself. How might satellite sensors contribute to C^3I in the TMD environment?

Consider the notional missile launch depicted in Figure 1.3. A satellite sensor in position to view a boosting TBM[21] can in principle provide useful information to a variety of theater defense platforms. By gathering information on the TBM trajectory, for example, a "forward track" of the missile can be derived, from which the time and location of missile impact can be estimated. If relayed to the target area in a timely manner, appropriate passive defensive measures may be employed. In addition, the forward track can include estimates of the missile position as a function of time along the trajectory. Such estimates could be used to cue search radars of active defense systems, and perhaps provide fire-control quality "launch baskets" for TBM interceptors.[22]

Similarly, a TBM "backtrack" to the launch point provided by satellite sensors could support attack operations with aircraft or ground-launched munitions. During the Gulf War, Scud launchers could be moved within minutes of missile firing, and after 15 minutes, could

---

[20]See D. Vaughan et al., *Evaluation of Operational Concepts for Countering Theater Ballistic Missiles*, Santa Monica, Calif.: RAND, WP-108, 1994.

[21]To simplify our discussion, we use the term "satellite sensor" to represent a spaceborne platform capable of detecting missiles during the boost-phase only. Sensors capable of detecting TBMs after booster burnout (e.g., Brilliant Eyes-type systems) are not considered here.

[22]In the case of boost-phase/ascent-phase intercept, time constraints may limit the utility of satellite-based information.

RAND*MR737-1.3*

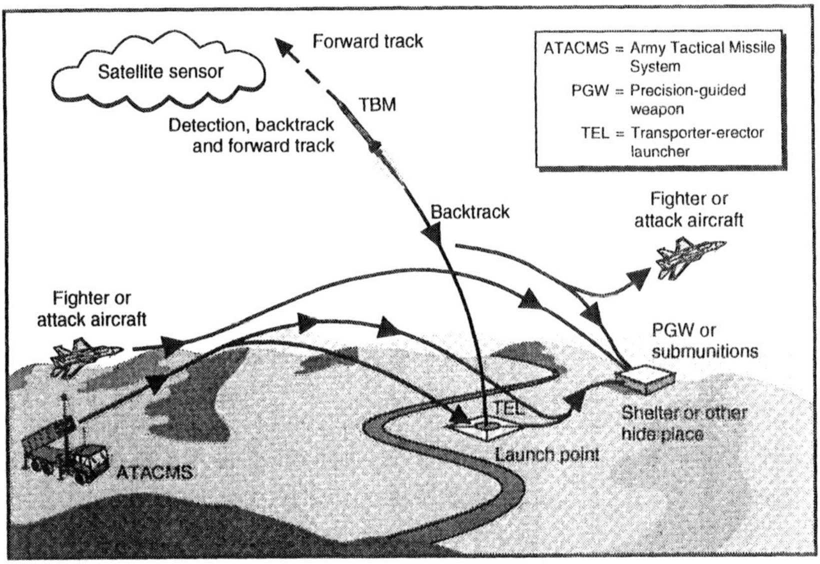

**Figure 1.3—Satellite Sensors Support Both Forward and
Target Area Defenses**

be anywhere within nine miles of the launch point, underscoring the importance of timely response.[23] By detecting and tracking the TBM during boost-phase,[24] however, the spaceborne systems considered here have the potential to supply information for such a response, and to do so nearly globally on an essentially continuous coverage basis.

ORGANIZATION OF THE REPORT

This report describes the operational implications of an established analytical procedure which, applied to notional satellite measurements, supplies information to a battle management function central

---

[23]Secretary of Defense, 1992, p. 224.

[24]Depending on the type of missile, boost-phases typically last between 30 and 120 sec. See Congressional Budget Office, 1994, p. 5.

to the theater missile defense mission. Chapter Two describes the theoretical underpinnings of the approach, known as linear filtering. The equations of a Kalman filter optimized for random measurement errors are derived for both linear systems and nonlinear systems in the linear approximation. The latter are applied to a notional TBM launch against Israel in Chapter Three, with an emphasis on analyzing launch point uncertainty and missile location uncertainty. Chapter Four discusses the effect of measurement bias on this filter, on a filter optimized for both random and bias errors, and on the trajectory analysis in both cases. Finally, Chapter Five offers some concluding remarks.

Chapter Two

THEORETICAL UNDERPINNINGS

This chapter briefly describes the formalism of a Kalman filter optimized for random errors. Equations are derived for both linear systems and nonlinear systems in the linear approximation.

LINEAR ESTIMATION AND PREDICTION

Consider a physical system whose characteristics may be fully determined at any time by the state of the system, x.[1] For a dynamical system, such a vector might contain the position, orientation, time, velocity, acceleration, and/or any other parameters relevant to describing its state. If measurements on such a system (in the absence of errors) yield observations that are proportional to the state vector (in the matrix sense), the system will obey the linear relation

$$z = Hx + v, \qquad (2.1)$$

where

z = p-dimensional measurement vector,

x = n-dimensional state vector of system,

H = a known $(p \times n)$-dimensional matrix,

v = measurement errors in z (p-dimensional). $\qquad$ (2.2)

---

[1]Our notation is similar to that of A. E. Bryson and Y.-C. Ho, *Applied Optimal Control,* New York: Hemisphere Publishing Corporation, 1975.

We assume the measurement errors are random, with vanishing expected value:

$$E(v) = 0. \tag{2.3}$$

Denote the estimate of the state before measurement by $\bar{x}$, and define the error covariance of the measurement and error covariance of the state before measurement by

$$R \equiv E\left(vv^T\right) \tag{2.4}$$

and

$$M \equiv E\left[(x - \bar{x})(x - \bar{x})^T\right], \tag{2.5}$$

respectively. Assuming x and v to be independent vectors obeying gaussian statistics, the probability density $p(x, v)$ is proportional to $\exp(-J)$, where J is the quadratic form

$$J = \frac{1}{2}\left[(x - \bar{x})^T M^{-1}(x - \bar{x}) + (z - Hx)^T R^{-1}(z - Hx)\right]. \tag{2.6}$$

One can show that J is minimized by the vector

$$\hat{x} = \bar{x} + PH^T R^{-1}(z - H\bar{x}) , \tag{2.7}$$

with P the error covariance of the state after measurement:

$$P \equiv E\left[(\hat{x} - x)(\hat{x} - x)^T\right]. \tag{2.8}$$

As a result, $x = \hat{x}$ and $v = \hat{v} \equiv z - H\hat{x}$ represent the "maximum likelihood estimate" of the state vector, in that they maximize $p(x, v)$ given the measurement z.[2] In other words, $\hat{x}$ is the most likely state vector resulting in the measurement z, given the statistical properties of x and v.[3]

---

[2]Ibid., p. 357.

[3]See A. Gelb (ed.), *Applied Optimal Estimation*, Reading, Massachusetts: The Analytic Sciences Corporation, 1974, p. 103.

It is a straightforward exercise to verify that P satisfies

$$P = \left(M^{-1} + H^T R^{-1} H \right)^{-1}$$

$$= M - MH^T \left(HMH^T + R \right)^{-1} HM. \tag{2.9}$$

If the state vector is of greater dimension than the measurement vector (i.e., if $n > p$), P is more easily obtained from the latter of the above equations. Note that this equation also predicts that measurements decrease the uncertainty in our knowledge of the state (i.e., $x^T P x \leq x^T M x$ for all n-dimensional vectors x), since the quantity subtracted from M above is nonnegative definite.

As alluded to above, the temporal evolution of this system may be accounted for by directly incorporating the time variable into the state vector. (This is a convenient choice when measurements are made continuously, as in radar tracking.) Alternatively, a discrete set of measurements occurring at different times may be accounted for by explicitly carrying a time index on the matrices and vectors composing the linear system. Let

$$z_i = H_i x_i + v_i \ , i = 1, \ldots, N \ , \tag{2.10}$$

where we assume

$$E(v_i) = 0 \ , \tag{2.11}$$

and the index i is an explicit time label for the sequence of measurements $1, \ldots, N$. We further assume that measurements at different times are uncorrelated; that is,

$$E(v_i v_j^{\ T}) = R_i \delta_{ij} \ , \tag{2.12}$$

where δ_{ij} is the Kronecker delta.[4] Generalizing Eq. (2.7), we write

$$\hat{x}_i = \bar{x}_i + P_i H_i^{\ T} R_i^{\ -1} (z_i - H_i \bar{x}_i) \ , \tag{2.13}$$

---

[4] $\delta_{ij} = 0$ for $i \neq j$; $\delta_{ij} = 1$ for $i = j$.

where sequential estimates are linked via

$$\bar{x}_{i+1} = \hat{x}_i \, , \qquad\qquad (\bar{x}_1 \text{ is given}) \qquad\qquad (2.14)$$

$$P_i = \left(M_i^{-1} + H_i^T R_i^{-1} H_i \right)^{-1}$$

$$= M_i - M_i H_i^T \left(H_i M_i H_i^T + R_i \right)^{-1} H_i M_i \, , \qquad\qquad (2.15)$$

and

$$M_{i+1} = P_i \, , \qquad\qquad (M_1 \text{ is given}). \qquad\qquad (2.16)$$

The above set of equations constitutes a particular form of a Kalman filter.[5] Equations (2.13)–(2.16) may be used to refine an initial estimate of the state ($\bar{x}_1$) and its corresponding error covariance (M_1) through the use of information obtained through the measurement process. The estimation sequence is represented in Table 2.1. Note that the matrices H_i and R_i must be specified to run the filter.

As formulated, the filter is optimized for random errors, which are uncorrelated from measurement to measurement. In Chapter Four, we will investigate the effect of "bias" errors, which are correlated.

LINEAR APPROXIMATION TO NONLINEAR SYSTEMS

Few physical systems are linear in the sense of Eq. (2.1); most are described by the nonlinear equation

$$z = h(x) + v \, , \qquad\qquad (2.17)$$

---

[5]R. E. Kalman, "A New Approach to Linear Filtering and Prediction," *Trans. ASME,* Vol. 82D, 1960, p. 35. More generally, the state vector has a known transition matrix (Φ), a known process noise distribution matrix (Γ), and is affected by a random process noise vector (w):

$$x_{i+1} = \Phi_i x_i + \Gamma_i w_i \, .$$

Since there are no disturbances to the state in our formulation (i.e., no process noise), we may choose our state vector to comprise initial value data, in which case Φ becomes the identity matrix. With this choice, the dynamics of the physical system are manifested in the measurement process and captured mathematically in the definition of the H -matrix. See A. E. Bryson and Y.-C. Ho, 1975, pp. 359–361.

Table 2.1

Estimation Sequence

Before Measurement	After Measurement
$\bar{x}_1, M_1$	$\hat{x}_1, P_1$
$\bar{x}_2 = \hat{x}_1, M_2 = P_1$	$\hat{x}_2, P_2$
$\bar{x}_3 = \hat{x}_2, M_3 = P_2$	$\hat{x}_3, P_3$
...	...
...	...
...	...

where h is a differentiable function of x. In the event that sufficient a priori knowledge of the state vector is obtainable, Eq. (2.17) can be expanded in a Taylor series about an initial estimate of the state. Denote this estimate by $\tilde{x}$, and the measurement to which it corresponds by $\tilde{z}$. Expanding about this estimate to linear order, one obtains

$$z - \tilde{z} \approx \frac{\partial z}{\partial x}\bigg|_{x=\tilde{x}} (x - \tilde{x}) + v \equiv H(x - \tilde{x}) + v . \qquad (2.18)$$

As a result, the developments of the preceding section can be applied if we simply shift the state and measurement vectors by appropriate constant vectors. Clearly, the matrices M, P, and R are unaffected by this redefinition [see Eqs. (2.4), (2.5), and (2.8)]. The estimate of the state, on the other hand, will be given by

$$\hat{x} = \bar{x} + PH^T R^{-1}(z - \tilde{z} - H(\bar{x} - \tilde{x})) , \qquad (2.19)$$

which reduces to

$$\hat{x} = \tilde{x} + PH^T R^{-1}(z - \tilde{z}) \qquad (2.20)$$

if we identify $\bar{x} \equiv \tilde{x}$.

In the linear approximation, then, it is easy to verify that the time evolution of the filter is governed by

$$\hat{x}_i = \bar{x}_i + P_i H_i^T R_i^{-1}(z_i - \tilde{z}_i - H_i(\bar{x}_i - \tilde{x})) , \qquad (2.21)$$

where

$$H_i \equiv \left. \frac{\partial z_i}{\partial x} \right|_{x=\tilde{x}} , \qquad (2.22)$$

and sequential estimates are linked via

$$\bar{x}_{i+1} = \hat{x}_i , \qquad\qquad (\bar{x}_1 \equiv \tilde{x} \text{ is given}) \qquad (2.23)$$

$$\begin{aligned} P_i &= \left(M_i^{-1} + H_i^T R_i^{-1} H_i \right)^{-1} \\ &= M_i - M_i H_i^T \left(H_i M_i H_i^T + R_i \right)^{-1} H_i M_i , \end{aligned} \qquad (2.24)$$

and

$$M_{i+1} = P_i , \qquad\qquad (M_1 \text{ is given}).[6] \qquad (2.25)$$

Finally, rewriting Eq. (2.18) in the sequential form

$$z_i - \tilde{z}_i = H_i(x - \tilde{x}) + v_i , \qquad (2.26)$$

where $\tilde{z}_i$ is the measurement vector at time index i corresponding to the expansion vector $\tilde{x}$, and defining

$$e_i \equiv \hat{x}_i - x , \qquad (2.27)$$

Eq. (2.21) can be rewritten as

$$e_i = (I - K_i H_i)e_{i-1} + K_i v_i , \qquad (2.28)$$

where the Kalman gain matrix is defined as

$$K_i \equiv P_i H_i^T R_i^{-1} . \qquad (2.29)$$

---

[6]Although not implemented in the present work, it is also possible to use the state vector estimate after measurement ($\hat{x}$) to update the expansion point ($\tilde{x}$), iterating until convergence is achieved.

As a result, since the covariance at the i*th* stage is given by

$$P_i \equiv E\left[e_i e_i^T\right],$$

(2.30)

one can show that the error in the state vector (e_i) and the estimate $\hat{x}_i$ are uncorrelated. In Chapter Three, we will apply this formulation to the problem of estimating and predicting theater ballistic missile trajectories.

ESTIMATION AND PREDICTION OF BALLISTIC MISSILE TRAJECTORIES

In this chapter, we apply the foregoing discussion to the case of ballistic missile trajectories. We begin with a description of the missile-sensor engagement.

GEOMETRY OF MISSILE-SENSOR ENGAGEMENT

As depicted in Figure 3.1, our notional sensor spins clockwise (i.e., in the right-handed sense) about an axis originating at the center of the earth and extending outward through the equator. Such a geometry may be used to describe satellite viewing from geosynchronous orbits.

To model the measurement process, it is useful to erect a coordinate system moving with the notional sensor. Consider first a spherical coordinate system centered on the earth, as illustrated in Figure 3.2. In these coordinates, the sensor location is described by a position vector with components (r, Θ, Φ). As usual, the spherical system is related to ordinary Cartesian coordinates through the transformation

$$
\begin{aligned}
x &= r \sin \Theta \cos \Phi \\
y &= r \sin \Theta \sin \Phi \\
z &= r \cos \Theta \, .
\end{aligned}
\tag{3.1}
$$

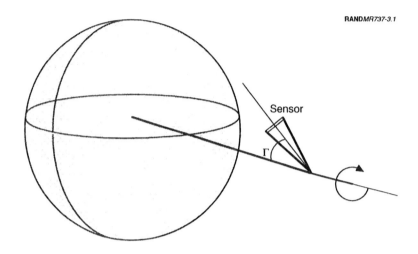

RAND*MR737-3.1*

Figure 3.1—Notional Sensor in Geosynchronous Orbit

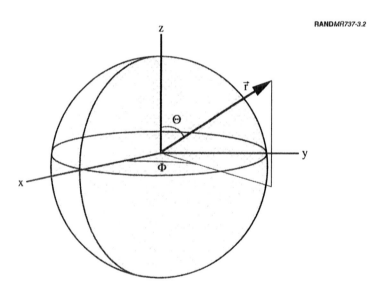

RAND*MR737-3.2*

Figure 3.2—Earth-Centered Coordinates

To transform this system into one rotating with the sensor, we make a sequence of coordinate transformations. First, translate the origin of the Cartesian system a vector amount $\vec{r}$. Next, rotate the (x, y, z) system an amount Φ about the z-axis (see Figure 3.3) using the matrix relation

$$\begin{pmatrix} x' \\ y' \\ z' \end{pmatrix} = \begin{pmatrix} \cos\Phi & \sin\Phi & 0 \\ -\sin\Phi & \cos\Phi & 0 \\ 0 & 0 & 1 \end{pmatrix}\begin{pmatrix} x \\ y \\ z \end{pmatrix}. \tag{3.2}$$

Now orient the x'-axis with the radial direction by rotating an amount $\Theta - \pi/2$ about the y'-axis:

$$\begin{pmatrix} x'' \\ y'' \\ z'' \end{pmatrix} = \begin{pmatrix} \sin\Theta & 0 & \cos\Theta \\ 0 & 1 & 0 \\ -\cos\Theta & 0 & \sin\Theta \end{pmatrix}\begin{pmatrix} x' \\ y' \\ z' \end{pmatrix}. \tag{3.3}$$

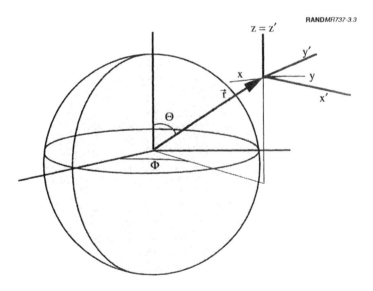

Figure 3.3—Rotated Coordinates

Next, tilt downward an amount Γ, again about the y'- or y''-axes:

$$\begin{pmatrix} x''' \\ y''' \\ z''' \end{pmatrix} = \begin{pmatrix} \cos \Gamma & 0 & -\sin \Gamma \\ 0 & 1 & 0 \\ \sin \Gamma & 0 & \cos \Gamma \end{pmatrix} \begin{pmatrix} x'' \\ y'' \\ z'' \end{pmatrix}. \tag{3.4}$$

The result is depicted in Figure 3.4.

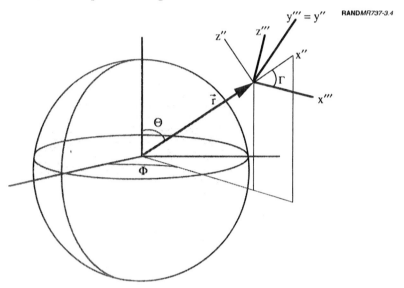

Figure 3.4—Tilted Coordinates

As the sensor spins, a vector in a coordinate system whose origin lies at the sensor position rotates counterclockwise (i.e., in the left-handed sense) with respect to the spin axis (see Figure 3.5). In terms of the coordinates above, the rotating vector $\vec{\rho}$ obeys[1]

$$\vec{\rho} = \vec{r}''' \cos \omega t + \vec{n}(\vec{n} \cdot \vec{r}''')(1 - \cos \omega t) + (\vec{r}''' \times \vec{n}) \sin \omega t , \tag{3.5}$$

where $\vec{n}$ defines the axis of (counterclockwise) rotation.

---

[1]H. Goldstein, *Classical Mechanics*, Reading, Massachusetts: Addison-Wesley, 1980, p. 164. Remember that a clockwise rotation of the coordinate system appears as a counterclockwise rotation of the vector.

As a result, applying Eqs. (3.1)–(3.5), we obtain the following relations for the position of an object viewed in a coordinate system that rotates as in Figure 3.5:

$$
\begin{aligned}
X &= x''' \cos \omega t + \left(x''' \cos \Gamma + z''' \sin \Gamma\right)\cos \Gamma\left(1 - \cos \omega t\right) \\
&\quad -y''' \sin \Gamma \sin \omega t \\
Y &= y''' \cos \omega t + \left(x''' \sin \Gamma - z''' \cos \Gamma\right)\sin \omega t \\
Z &= z''' \cos \omega t + \left(x''' \cos \Gamma + z''' \sin \Gamma\right)\sin \Gamma\left(1 - \cos \omega t\right) \\
&\quad +y''' \cos \Gamma \sin \omega t ,
\end{aligned}
\tag{3.6}
$$

where

$$
\begin{aligned}
x''' &= x'' \cos \Gamma - z'' \sin \Gamma \\
y''' &= y'' \\
z''' &= x'' \sin \Gamma + z'' \cos \Gamma
\end{aligned}
\tag{3.7}
$$

and

$$
\begin{aligned}
x'' &= x \sin \Theta \cos \Phi + y \sin \Theta \sin \Phi + z \cos \Theta \\
y'' &= -x \sin \Phi + y \cos \Phi \\
z'' &= -x \cos \Theta \cos \Phi - y \cos \Theta \sin \Phi + z \sin \Theta \cdot
\end{aligned}
\tag{3.8}
$$

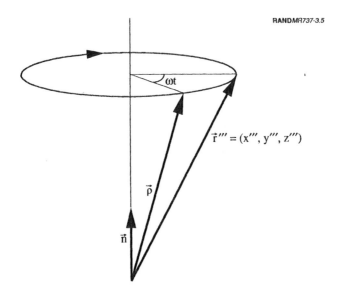

RANDMR737-3.5

ωt

$\vec{r}''' = (x''', y''', z''')$

$\vec{\rho}$

$\vec{n}$

Figure 3.5—Spinning Coordinates

Thus, in addition to the positions of the sensor and missile, the rotation rate and tilt of the sensor need to be specified to adequately describe the engagement in this formalism.

With Equation (3.6) at hand, define the angles

$$\alpha_v \equiv \tan^{-1}\left(-\frac{Z}{X}\right) \tag{3.9}$$

and

$$\alpha_h \equiv \tan^{-1}\left(-\frac{Y}{X}\right), \tag{3.10}$$

representing "vertical" and "horizontal" angles in the sensor's frame, respectively. [The boresight of the sensor points toward the earth along a ray through the origin of the Y–Z coordinate plane (i.e., through $\alpha_v = \alpha_h = 0$)]. When an object passes through the sensor's field of view, $\alpha_h = 0$. We may therefore use the rotation phase angle $\Omega \equiv \omega t$ to define another angle representing the rotational position of the sensor when the object passes by.[2] Setting Eq. (3.10) to zero, and using Eq. (3.6), we obtain

$$\Omega = \tan^{-1}\left(\frac{y'''}{z''' \cos \Gamma - x''' \sin \Gamma}\right). \tag{3.11}$$

(Loosely speaking, Ω represents the angular position of a hand on a clock, where the face represents the disk of the earth as seen from the position of the sensor.) In terms of the filter analysis, the two sets of angles (α_v, α_h and α_v, Ω) are equivalent.[3] Unless otherwise stated, we will assume

---

[2]By defining a measurement to occur when $\alpha_h = 0$, the sensor is more properly described as a vertical slit with no horizontal extent. Note that the slit is aligned toward the north when $t = 0$.

[3]That is, measurement vectors taken as

$$\begin{pmatrix} z_1 \\ z_2 \end{pmatrix} = \begin{pmatrix} \alpha_v \\ \Omega \end{pmatrix} \text{ or } \begin{pmatrix} z_1' \\ z_2' \end{pmatrix} = \begin{pmatrix} \alpha_v \\ \alpha_h \end{pmatrix}$$

yield the same results. Note, however, that by mathematical convention the $\tan^{-1}$ function assumes values between –90 and +90 degrees, and so cannot represent angles

$$\begin{pmatrix} z_1 \\ z_2 \end{pmatrix} = \begin{pmatrix} \alpha_v \\ \Omega \end{pmatrix}. \tag{3.12}$$

It is convenient to parameterize the error in Ω in terms of the error in α_h. Differentiating Eq. (3.10) (holding the missile position constant), setting $\alpha_h = 0$, and defining the quantity

$$\begin{aligned} D = \cos\Omega + \left(\cos\Gamma + \frac{z'''}{x'''} \sin\Gamma \right) \cos\Gamma(1 - \cos\Omega) \\ - \frac{y'''}{x'''} \sin\Gamma \sin\Omega, \end{aligned} \tag{3.13}$$

we find

$$\delta\alpha_h = \frac{\delta\Omega}{D}\left[\frac{y'''}{x'''} \sin\Omega - \left(\sin\Gamma - \frac{z'''}{x'''} \cos\Gamma \right) \cos\Omega \right]. \tag{3.14}$$

With a little algebra, we may rewrite the above as

$$\delta\alpha_h = -\delta\Omega\left(\sin\Gamma + \tan\alpha_v \cos\Gamma \right). \tag{3.15}$$

FILTER METHODOLOGY

To calculate the H-matrix, we define a template for a given missile from its range-altitude data, which are obtained by modeling the missile's flight in the atmosphere of a spherical, nonrotating earth.[4] This trajectory is used as a baseline from which perturbations—and ultimately, the H-matrix elements—are generated. In the field, sensor measurements would be obtained from the actual missile under observation; here, to simply estimate the errors one might expect using the filter technique (as opposed to estimating the state vector),

---

in the second and third quadrants. In applying Eq. (3.10), this is not a problem for most practical geometries because X is usually negative. Such is not the case for Eq. (3.11), so that special care must be taken in applying this equation. One solution is to use Eq. (3.10) to determine when a measurement occurs (i.e., when $\alpha_h = 0$), and then substitute the relevant coordinates [using Eq. (3.6)] into Eq. (3.11) to find Ω.

[4]For intermediate- and shorter-range missiles, neglecting rotational effects is usually justifiable.

we simulate this process by taking the measurements on the template trajectory.

The (constant) state vector is defined by

$$
x \equiv \begin{pmatrix} x_1 \\ x_2 \\ x_3 \\ x_4 \\ x_5 \\ x_6 \end{pmatrix},
\tag{3.16}
$$

where

x_1 = launch latitude,

x_2 = launch longitude,

x_3 = launch heading,

x_4 = launch time,

x_5 = launch altitude,

x_6 = loft angle characterizing missile pitch-over.[5]

From Eq. (2.18), we can obtain the H-matrix by perturbing the state vector elements and examining the changes on the measurement vector z. In this manner, for the case of a two-dimensional measurement vector [i.e., with components (z_1, z_2)], the elements of H are given by

$$
H = \begin{pmatrix} \dfrac{\partial z_1}{\partial x_1} & \dfrac{\partial z_1}{\partial x_2} & \dfrac{\partial z_1}{\partial x_3} & \dfrac{\partial z_1}{\partial x_4} & \dfrac{\partial z_1}{\partial x_5} & \dfrac{\partial z_1}{\partial x_6} \\ \dfrac{\partial z_2}{\partial x_1} & \dfrac{\partial z_2}{\partial x_2} & \dfrac{\partial z_2}{\partial x_3} & \dfrac{\partial z_2}{\partial x_4} & \dfrac{\partial z_2}{\partial x_5} & \dfrac{\partial z_2}{\partial x_6} \end{pmatrix} \Bigg|_{x=\tilde{x}},
\tag{3.17}
$$

---

[5]This parameterizes a family of templates based on early pitch-over followed by zero angle of attack for the remainder of the flight trajectory. More generally, we could have assumed any family of flight trajectory templates for which variation of the loft angle is determined by a one-parameter family of steering functions applied early in the boost-phase.

where partial derivatives with respect to one state vector element are calculated holding other elements constant and evaluated at the initial guess $x = \tilde{x}$.[6] Variations in launch position (x_1, x_2) and heading (x_3) are straightforward, simply changing the geometric relationship between missile and sensor. Variations in launch time (x_4), on the other hand, move the position of the missile forward or backward in its time history. If one were to imagine a sequence of beads on a wire representing points along the trajectory (see Figure 3.6), such variations could be described by sliding the beads backward or forward along the wire. (The "beads" shown in Figure 3.6 represent trajectory points plotted every five seconds.) Varying launch altitude (x_5) causes more than a vertical translation of the trajectory, since drag depends on the atmospheric density, an approximately exponential function of altitude. Finally, to allow for planar variations in the trajectory template, we vary the loft angle (x_6) during the pitch-over

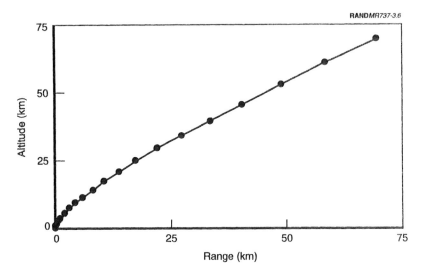

Figure 3.6—Boost-Phase of Notional Missile

---

[6]The complexity of this problem demands that these derivatives be calculated numerically.

phase of flight. Loft angle is identical to the angle of attack, defined with respect to the instantaneous velocity vector of the missile. At launch and prior to pitch-over, the missile velocity is in the vertical direction (defined locally). As x_6 increases, roughly speaking, the vertical speed of the missile is converted to horizontal speed, so that small loft angles result in lofted trajectories and large loft angles cause trajectories to depress.

In what follows, we consider the estimation/prediction problem for the case of a notional TBM whose trajectory is depicted in Figure 3.7. As the figure illustrates, this missile has a boost-phase of 100-sec duration and a total range of 1200 km. For an initial guess, we assume a launch in Iran (at 34.01° latitude, 47.40° longitude) with a 263° heading,[7] impacting Tel Aviv at 32.05° latitude, 34.77° longitude. The relevant geometry is illustrated in Figure 3.8, where the satellites

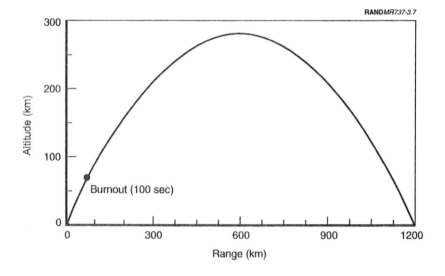

Figure 3.7—Notional 100-sec-Burn Missile Trajectory

---

[7]0° represents due north and 90°, due east.

RAND*MR737-3.8*

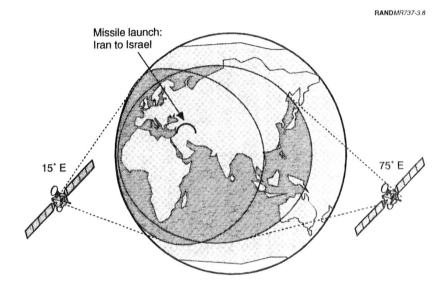

Figure 3.8—Geometry of TBM Trajectory and Sensors

are positioned in geosynchronous orbit at $0°$ latitude, and $15°$ and $75°$ east longitude, respectively.[8]   (The equations describing trajectories on a spherical earth may be found in the Appendix.)

As depicted in Figure 3.9, the two satellites independently sample the missile boost-phase, each measuring two angles (z_1, z_2) at 20-sec intervals (the assumed revisit time[9]). Since the notional TBM takes 42 sec to reach an altitude of 10 km, a sensor unable to see through a cloud layer at this altitude would, roughly speaking, be denied two

---

[8]Geosynchronous orbit about a spherical earth occurs at an altitude of roughly 35,800 km as measured from the equatorial surface (equivalent to a radius vector about 42,200 km in extent as measured from the center of the earth). For satellites at $0°$ latitude, the disk of the earth subtends a half-angle of roughly $8.8°$, so that a $4.4°$ tilt angle with a $4.4°$ field of view covers the disk completely as the sensor revolves about its spin axis.

[9]In a sensitivity excursion, we later examine the effects on the trajectory analysis of varying the revisit time.

RAND*MR737-3.9*

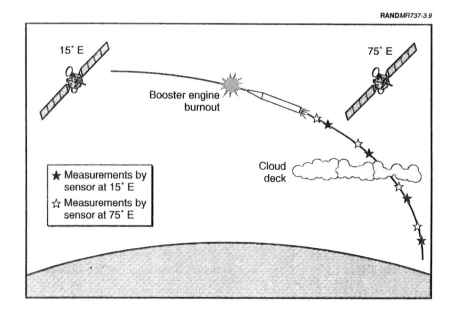

Figure 3.9—Boost-Phase Measurement Sequence

measurements. For simplicity, we assume clouds do not present such a problem, and ignore effects of early booster engine cutoff.[10]

First, consider the satellite positioned at 75° longitude. The H-matrix corresponding to each measurement may be calculated numerically, utilizing Eqs. (3.1)–(3.10) and (3.12). For the example at hand, we approximate derivatives by differences using a step

$$\delta x = \begin{pmatrix} 0.01 \\ 0.01 \\ 0.01 \\ 1 \\ 0.01 \\ 0.01 \end{pmatrix}, \qquad (3.18)$$

---

[10]That is, we assume full-burn trajectories throughout. Note, however, that the time at burnout is still uncertain, owing to an uncertainty in the missile launch time.

with angles measured in degrees, time in seconds, and altitude in kilometers. By varying the state vector elements independently by the amounts illustrated above, the change in the measured angles may be calculated and derivatives determined. Five measurements occur during boost-phase in this example, at roughly 18, 38, 58, 78, and 98 sec after launch. (The 20-sec periodicity reflects the revisit time of the sensor. The motion of the missile is negligible here because it is viewed from geosynchronous altitude.) Note that at 18 sec, the H-matrix reads

$$H = \begin{pmatrix} 8.0\times10^{-2} & -6.2\times10^{-2} & 5.4\times10^{-6} & -2.7\times10^{-4} & 1.2\times10^{-3} & 2.3\times10^{-5} \\ 1.0 & 8.9\times10^{-1} & 5.2\times10^{-5} & 7.0\times10^{-4} & 3.4\times10^{-5} & -8.0\times10^{-4} \end{pmatrix},$$

$$(3.19)$$

whereas just prior to burnout (98 sec), it is given by

$$H = \begin{pmatrix} 8.0\times10^{-2} & -6.3\times10^{-2} & 9.0\times10^{-4} & -3.4\times10^{-3} & 4.3\times10^{-3} & -4.1\times10^{-3} \\ 1.0 & 8.7\times10^{-1} & 8.9\times10^{-3} & 2.4\times10^{-2} & 8.4\times10^{-3} & -9.7\times10^{-2} \end{pmatrix}.$$

$$(3.20)$$

This illustrates that the H-matrix is time-dependent.

One would not expect matrix elements corresponding to changes in launch position (i.e., the first two columns of H) to vary appreciably during the missile flight, since, in effect, these changes amount to sliding the "wire" trajectory as a whole over the earth's surface. We may estimate these using order-of-magnitude approximations to the missile-sensor engagement:

$$\left\{ \begin{array}{l} R_{GEO} \cdot \Delta\alpha_v \sim R_{EARTH} \cdot \Delta l \\ R_{GEO} \cdot \Delta\alpha_v \sim R_{EARTH} \cdot \Delta L \end{array} \right\} \Rightarrow \frac{\partial z_1}{\partial x_1}, \frac{\partial z_1}{\partial x_2} \sim \frac{R_{EARTH}}{R_{GEO}} \sim 10^{-1}, \quad (3.21)$$

and

$$\left\{ \begin{array}{l} R_{EARTH} \cdot \Delta\Omega \sim R_{EARTH} \cdot \Delta l \\ R_{EARTH} \cdot \Delta\Omega \sim R_{EARTH} \cdot \Delta L \end{array} \right\} \Rightarrow \frac{\partial z_2}{\partial x_1}, \frac{\partial z_2}{\partial x_2} \sim 1, \quad (3.22)$$

with l the latitude (x_1), L the longitude (x_2), R_{EARTH} the earth's radius, and R_{GEO} the geosynchronous altitude. [Recall that (z_1, z_2) =

(α_v, Ω).] Similar scaling arguments may be constructed for other matrix elements, although these exhibit a more complicated geometric and temporal dependence.[11]

Once H is calculated, the covariance matrix of random errors R is constructed. Because we choose to define $\delta\Omega$ in terms of $\delta\alpha_h$ [see Eq. (3.15)], R will exhibit a slight time-dependence, as the following illustrates: For a 30-microradian random error, the R-matrix reads

$$R = \begin{pmatrix} 3.0\times10^{-6} & 0 \\ 0 & 2.3\times10^{-4} \end{pmatrix} \quad \text{at 18 sec, whereas} \quad (3.23)$$

$$R = \begin{pmatrix} 3.0 \times 10^{-6} & 0 \\ 0 & 2.2 \times 10^{-4} \end{pmatrix} \quad \text{at 98 sec.} \quad (3.24)$$

By specifying an initial guess for P [i.e., M_1—see Eq. (2.25)], we may run the filter algorithm. We next describe some notional results.

NOTIONAL RESULTS

Launch Point Uncertainty (LPU)

Determining the uncertainty associated with missile launch location is a useful example of the Kalman filter technique's utility. Describe the launch position with the vector

$$w \equiv \begin{pmatrix} w_1 \\ w_2 \end{pmatrix}, \quad (3.25)$$

where (w_1, w_2) are launch latitude and longitude, respectively. Writing the above as

$$w \equiv Fx, \quad (3.26)$$

---

[11]In particular, because the prevalent effects of loft angle variations are manifested later in the trajectory, the matrix elements corresponding to these variations vary by more than two orders of magnitude over the course of the boost-phase. Thus, measurements occurring early in the boost-phase are much less sensitive to these types of variations than measurements occurring near missile burnout.

where

$$F = \begin{pmatrix} 1 & 0 & 0 & 0 & 0 & 0 \\ 0 & 1 & 0 & 0 & 0 & 0 \end{pmatrix}, \tag{3.27}$$

it is easy to see that

$$W \equiv E\left[(w - \hat{w})(w - \hat{w})^T\right] = FPF^T = \begin{pmatrix} P_{11} & P_{12} \\ P_{21} & P_{22} \end{pmatrix}. \tag{3.28}$$

Launch point uncertainty is thus determined from a 2×2 submatrix composing P.

The probability that w lies within the ellipse

$$(w - \hat{w})^T W^{-1}(w - \hat{w}) = \ell^2 \tag{3.29}$$

is given by[12]

$$\int_0^\ell \exp\left(-\frac{r^2}{2}\right) r dr = 1 - \exp\left(-\frac{\ell^2}{2}\right), \tag{3.30}$$

or 0.393, 0.865, and 0.989 for $\ell = 1, 2$, and 3, respectively.

Consider the notional trajectory discussed previously (Figure 3.7). As an initial estimate of the covariance, we use

$$M = \begin{pmatrix} 1 & 0 & 0 & 0 & 0 & 0 \\ 0 & 1 & 0 & 0 & 0 & 0 \\ 0 & 0 & 400 & 0 & 0 & 0 \\ 0 & 0 & 0 & 400 & 0 & 0 \\ 0 & 0 & 0 & 0 & 1 & 0 \\ 0 & 0 & 0 & 0 & 0 & 1 \end{pmatrix}, \tag{3.31}$$

corresponding (at the one-sigma level) to a $1°$ uncertainty in launch latitude and longitude, $20°$ uncertainty in launch heading, 20-sec uncertainty in launch time, 1-km uncertainty in launch altitude, and a $1°$ uncertainty in loft angle. For a single satellite positioned at $0°$ latitude, $75°$ longitude, Figure 3.10 illustrates the $\ell = 2$ launch point un-

---

[12]A. E. Bryson and Y.-C. Ho, 1975, pp. 310–311.

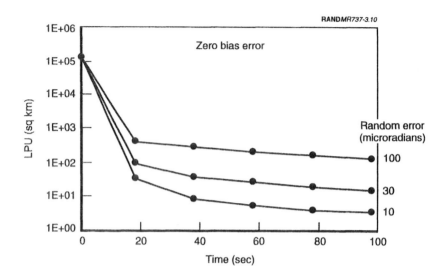

Figure 3.10—Sensitivity of LPU ($\ell = 2$) to Random Error (One Sensor)

certainty as a function of time for various random errors (100, 30, and 10 microradians, respectively).[13]

In the absence of measurement errors, six angles would uniquely determine the six-dimensional state vector we are estimating (assuming our template is exact). Since each measurement provides two angles, only three measurements would be required to specify the state. Consequently, as is clear in all cases above, a priori uncertainties are reduced most rapidly by the first few measurement and at a slower pace thereafter.

Next consider a second satellite positioned at 0° latitude, 15° longitude. Processing its measurements sequentially with those of the first satellites (i.e., stereo processing) may significantly reduce the launch point uncertainty. Figure 3.11 depicts the LPU for this case, derived after the last measurement has been made. The LPU calculated monoscopically from each separate sensor is also shown, indi-

---

[13]We assume $\delta\alpha_h = \delta\alpha_v$ for simplicity [see Eqs. (3.9)–(3.15)].

RANDMR737-3.11

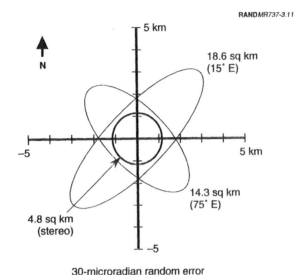

30-microradian random error

Figure 3.11—LPUs (ℓ = 2) for Two Sensors with Random Errors

cating how a different viewing geometry may lead to different results.[14] In all cases, a 30-microradian random error is assumed.

Figure 3.12 illustrates the $\ell = 2$ launch point uncertainty as a function of time for various random errors (100, 30, and 10 microradians) in the case of stereo processing. (The second sensor provides measurements at roughly 2, 22, 42, 62, and 82 sec.) As expected on intuitive grounds, the LPU derived after the final measurement has been made scales roughly as the square of the random error.

Finally, consider a symmetric example, where the same missile is launched from 35° latitude, 45° longitude heading due north. If the sensors rotate in the opposite direction relative to each other, we would expect the symmetry of the problem to manifest itself in the results. As Figure 3.13 illustrates, this is indeed the case.

---

[14]As shown later, a perfectly symmetric example results in identical LPU values calculated from each sensor.

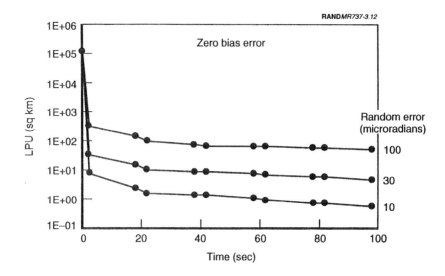

Figure 3.12—Sensitivity of LPU ($\ell = 2$) to Random Error (Two Sensors)

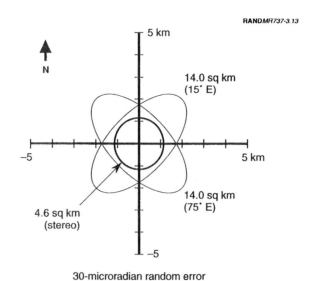

Figure 3.13—LPUs ($\ell = 2$) for a Symmetric Example

Missile Location Uncertainty (MLU)

Determining the uncertainty associated with missile location at any point along its trajectory is another useful application of the technique.[15] Describe the instantaneous missile position at time t by the vector

$$y(t) \equiv \begin{pmatrix} y_1(t) \\ y_2(t) \\ y_3(t) \end{pmatrix}, \tag{3.32}$$

where the elements (y_1, y_2, y_3) are referenced to a Cartesian coordinate system centered at the center of the earth. (We may choose the y_3-direction to intersect the poles, and the y_1–y_3 plane to intersect Greenwich.) Although y(t) is a nonlinear function of the state vector, we may expand to linear order about an initial estimate of the state [see Chapter Two, Eqs. (2.17)–(2.22)]. In similar fashion, we find

$$y - \tilde{y} \approx \left.\frac{\partial y}{\partial x}\right|_{x=\tilde{x}} (x - \tilde{x}) \equiv G(x - \tilde{x}). \tag{3.33}$$

The covariance of y(t) will therefore be given by

$$O \equiv E\left[(y - \hat{y})(y - \hat{y})^T\right] = GPG^T, \tag{3.34}$$

once the G-matrix is determined. (This may be accomplished numerically, using a procedure similar to that used in determining H.)

The probability that y lies within the ellipsoid

$$(y - \hat{y})^T O^{-1}(y - \hat{y}) = \ell^2 \tag{3.35}$$

is given by[16]

$$\sqrt{\frac{2}{\pi}} \int_0^\ell \exp\left(-\frac{r^2}{2}\right) r^2 dr = erf\left(\frac{\ell}{\sqrt{2}}\right) - \ell\sqrt{\frac{2}{\pi}} \exp\left(-\frac{\ell^2}{2}\right), \tag{3.36}$$

or 0.199, 0.739, and 0.971 for $\ell = 1$, 2, and 3, respectively.

---

[15]The mathematical framework developed for analyzing MLU could be applied more generally to other uncertainties—for example, missile velocity (in three dimensions), impact point (in two dimensions).

[16]A. E. Bryson and Y.-C. Ho, 1975.

Figure 3.14 depicts the missile location uncertainty as a function of time along the trajectory in the case of 100-, 30-, and 10-microradian random errors, respectively. As illustrated, the uncertainty volume increases monotonically until the latter part of the trajectory, when the MLU turns over.[17] (As a point of reference, a sphere of 62-km radius encloses a volume of roughly 10^6 km^3.) Results for two sensors processed stereoscopically are shown in Figure 3.15.

Revisit Time Sensitivities

In general, decreasing the revisit time allows more measurements to be made and, consequently, more information to be obtained about

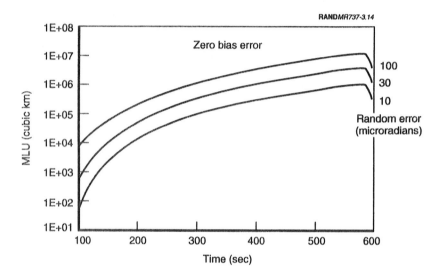

Figure 3.14—Sensitivity of MLU ($\ell = 2$) to Random Error (One Sensor)

---

[17]By examining the trajectories with perturbed launch times, altitude, and loft, one finds for the example at hand that the deviation from the nominal baseline trajectory begins to decrease at a reentry altitude of approximately 30,000 ft. This effect, manifested in the decreasing uncertainty 580 sec into the flight, is related to both the atmospheric degradation of the missile velocity upon reentry and our choosing a minimum-energy trajectory to perturb about.

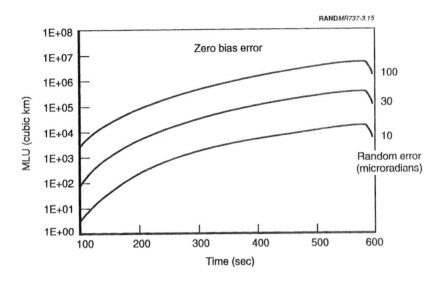

Figure 3.15—Sensitivity of MLU ($\ell = 2$) to Random Error (Two Sensors)

the missile trajectory. Figure 3.16 illustrates the LPU for various re-visit times, spanning the range of 2.5–40 sec. At late times, note that the LPU scales roughly linearly with the number of measurements.

Finally, Figure 3.17 shows the effect of revisit time on missile location uncertainty. As is evident from this plot, an order-of-magnitude reduction in revisit time generates more than an order-of-magnitude reduction in uncertainty volume.

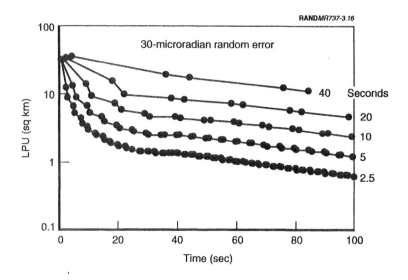

Figure 3.16—Sensitivity of LPU ($\ell = 2$) to Revisit Time (Two Sensors)

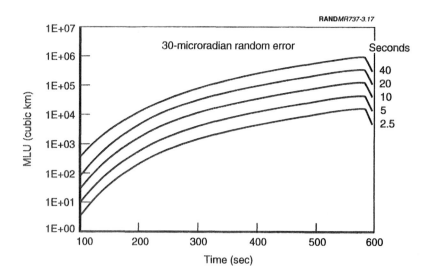

Figure 3.17—Sensitivity of MLU ($\ell = 2$) to Revisit Time (Two Sensors)

Chapter Four

THE EFFECT OF BIAS ERRORS

When correlation times[1] are shorter than the timescales of the measurement sequence, we may approximate the errors as uncorrelated, which we have done thus far in our treatment of a Kalman filter optimized for random errors. In many situations, however, bias errors—which are correlated from measurement to measurement—dominate the uncertainty. In such cases, one must explicitly account for their effects on the uncertainty analysis.

Consider the linear system

$$z_i - \tilde{z}_i = H_i(x - \tilde{x}) + v_i + b_i , \qquad (4.1)$$

where x, z_i, H_i, and v_i are as before (see Chapter Two), and

$$b_i = \text{bias error in measurement of } z_i \text{ (p-dimensional).} \qquad (4.2)$$

We assume that expected values of both the random and bias errors vanish (i.e., $E(v_i) = E(b_i) = 0$), but that bias terms exhibit correlations from measurement to measurement:

$$E(b_i b_j^T) \neq 0 . \qquad (4.3)$$

Although sensor biases may be eliminated (to the extent possible) by repeated calibration, some slowly varying sensor errors may remain.

---

[1]The correlation time, τ, measures roughly the mean time between two successive maxima (or minima) of some fluctuating function, f(t). As such, τ characterizes the rate at which f(t) varies. See F. Reif, *Fundamentals of Statistical and Thermal Physics*, New York: McGraw-Hill, 1965, p. 561.

If these are small enough, their effects may be insignificant or tolerable, although ignoring them in the filter design and error analysis procedure runs the risk of producing misleading uncertainty estimates. Here, two viewpoints are possible: (1) the estimate of the measurement error variance is correct, but it is not recognized that an element of it results from bias, or (2) the variance estimate is correct for the uncorrelated measurement errors, but an additional bias error is present. In any case, the effect of bias on a filter optimized for random errors alone can be examined.

However, if the bias errors are large enough, further calibration or redesign of the filter—wherein bias effects are modeled explicitly—may be desirable. In what follows, we examine the following two questions: What effect will the additional bias term (in Eq. 4.1) have on the existing filter optimized for random errors alone? How does this compare with results obtained using a filter designed to account optimally for both random and bias errors?

SUBOPTIMAL TREATMENT OF BIAS[2]

To address the first question, it is useful to adopt the nomenclature of linear systems analysis as developed at the end of Chapter Two.[3] At the i*th* stage of the filtering process, define the difference between the state of the system and its best estimate (after the i*th* measurement) as [(Eq. (2.27)]

$$e_i \equiv \hat{x}_i - x .$$ (4.4)

The linear system evolves according to

$$e_{i+1} = A_{i+1}e_i + K_{i+1}v_{i+1} + K_{i+1}b_{i+1} ,$$ (4.5)

---

[2]Here we examine the effect of bias errors on an existing filter optimized for random errors. There is, consequently, no reason to expect our results to be optimal. For this reason, as well as for brevity, in what follows we refer to this treatment as *suboptimal.*

It is natural to ask why one would bother using a suboptimal formulation. The answer is that if redesigning an existing filter optimized for random errors alone is not desirable, the suboptimal approach allows the effects of bias on that filter to be examined, albeit as an afterthought.

[3]For a similar but more general treatment of this problem, see B. Friedland, "Treatment of Bias in Recursive Filtering," *IEEE Transactions on Automatic Control,* Vol. AC-14, No. 4, August 1969, pp. 359–367.

where

$$A_i = I - K_i H_i$$
$$K_i = P_i^r H_i^T R_i^{-1}$$

(4.6)

and P_i^r is the error covariance predicted by the filter in the absence of bias [see Eq. (2.28)]. Explicitly writing the random and bias contributions to the error as

$$e_{i+1} \equiv e_{i+1}^r + e_{i+1}^b ,$$

(4.7)

we find

$$e_{i+1}^r = A_{i+1} e_i^r + K_{i+1} v_{i+1}$$
$$e_{i+1}^b = A_{i+1} e_i^b + K_{i+1} b_{i+1} .$$

(4.8)

To recover the notion of estimates before and after measurement (see Table 2.1), we specify an initial guess e_0^r whose square expected value is the covariance of the state before the first measurement; that is,

$$M_1 \equiv E\left[e_0^r \left(e_0^r \right)^T \right].$$

(4.9)

In this way, recursive application of Eq. (4.8) yields

$$e_1^r = A_1 e_0^r + K_1 v_1$$
$$e_2^r = A_2 e_1^r + K_2 v_2$$
$$e_3^r = A_3 e_2^r + K_3 v_3$$
$$\vdots$$
$$e_n^r = A_n e_{n-1}^r + K_n v_n .$$

(4.10)

Proceeding similarly for the bias contribution, one finds[4]

---

[4]Choosing $e_0^b = 0$ is equivalent to requiring the covariance of the state vector before the first measurement (i.e., M_1) to result entirely from random errors. Since these drive the filter, this choice is appropriate.

$$e_1^{\,b} = K_1 b_1$$
$$e_2^{\,b} = A_2 e_1^{\,b} + K_2 b_2$$
$$e_3^{\,b} = A_3 e_2^{\,b} + K_3 b_3 \qquad\qquad (4.11)$$
$$\vdots$$
$$e_n^{\,b} = A_n e_{n-1}^{\,b} + K_n b_n \quad.$$

Collecting terms for $b_i = b$ (a constant), one may write

$$e_n^{\,b} = \Big[K_n + A_n K_{n-1} + A_n A_{n-1} K_{n-2} + \cdots + A_n A_{n-1} A_{n-2} \cdots A_2 K_1 \Big] b$$

$$\equiv \Psi_n b \ ,$$

$$(4.12)$$

so that the effect of bias on the error covariance of the state will be given by

$$P_n^{\,b} \equiv E\left[e_n^{\,b} \left(e_n^{\,b} \right)^T \right] = \Psi_n E\left[bb^T \right] \Psi_n^{\,T} \ , \qquad (4.13)$$

where the total covariance of the state after the n*th* measurement is

$$P_n = P_n^{\,r} + P_n^{\,b} \ . \qquad\qquad (4.14)$$

Thus, specifying

$$B_i \equiv E(b_i b_i^{\,T}) = E(bb^T) \qquad\qquad (4.15)$$

enables one to account for (constant) bias effects. Finally, note that Ψ_i also time evolves linearly, as one might expect:

$$\Psi_{i+1} = A_{i+1} \Psi_i + K_{i+1} \ . \qquad\qquad (4.16)$$

In many applications, the bias will not be constant. Indeed, in our own formulation utilizing the measurement vector $(z_1, z_2) = (\alpha_v, \Omega)$, constant bias errors associated with (α_v, α_h) translate into time-dependent errors associated with Ω. In this case, however, we may modify the formalism in a straightforward way. Let

$$e_{i+1}^{\,b} = A_{i+1} e_i^{\,b} + K_{i+1} \sigma_{i+1} b_{i+1} \ , \qquad\qquad (4.17)$$

where

$$\sigma_i b_i \equiv \begin{pmatrix} 1 & 0 \\ 0 & g_i \end{pmatrix} \begin{pmatrix} b_i \\ b_i \end{pmatrix} = \begin{pmatrix} b_i \\ g_i b_i \end{pmatrix} \tag{4.18}$$

and g_i is a time-dependent factor. Then Eq. (4.16) is simply modified as

$$\Psi_{i+1} = A_{i+1}\Psi_i + K_{i+1}\sigma_{i+1} . \tag{4.19}$$

In the case of multiple sensors processed sequentially, we may further modify the formalism to allow for multiple biases. In the case of two sensors, write

$$e_{i+1}{}^b = A_{i+1}e_i{}^b + K_{i+1}\sigma_{i+1}b_{i+1} + K_{i+1}\sigma_{i+1}c_{i+1} , \tag{4.20}$$

where c_i is the bias of the additional sensor. Proceeding as above, we define

$$e_i{}^b \equiv \Psi_i b_i + X_i c_i . \tag{4.21}$$

Assume the measurements are processed in an alternating manner, with one sensor measuring at odd values of i and the other at even values. For odd i, the bias contributions will obey

$$\Psi_i = A_i\Psi_{i-1} + K_i\sigma_i$$
$$X_i = A_i X_{i-1} , \tag{4.22}$$

whereas for even i,

$$X_i = A_i X_{i-1} + K_i\sigma_i$$
$$\Psi_i = A_i\Psi_{i-1} . \tag{4.23}$$

Notional Results

For the example discussed in Chapter Three, the effect of bias on the existing filter is illustrated in Figure 4.1, where random and bias errors are *both* assumed to be 30 microradians. The launch point uncertainty ellipses obtained are larger than in Figure 3.11.

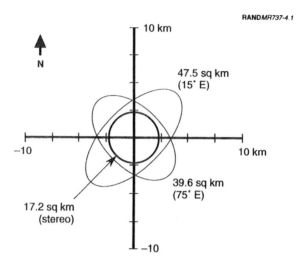

30-microradian random and bias errors

Figure 4.1—LPUs ($\ell = 2$) for Two Sensors with Random and Bias Errors (Suboptimal Filter)

Figure 4.2 illustrates the time evolution of the launch point estimate for the case of a 30-microradian random error and 100-, 30-, and 10-microradian bias errors, respectively. Unlike the case with random errors alone (Figure 3.12), the LPU derived after the final measurement does not scale as the square of the error. Moreover, the importance of bias is apparent in the large difference between the 30- and 100-microradian cases. The missile location uncertainty shows similar qualitative behavior (see Figure 4.3).

The effect of varying revisit time in the presence of random and bias errors is illustrated in Figures 4.4 and 4.5, respectively. In contrast to the unbiased case (Figure 3.16), Figure 4.4 indicates that the LPU derived after the final measurement is relatively insensitive to revisit time when both 30-microradian random and bias errors are present. In effect, a point is reached in the filtering sequence where additional measurements containing unknown bias errors provide information of limited utility. This underscores the point that *a high rate of data collection will not reduce the launch point uncertainty significantly unless random errors dominate the measurement process.* Indeed,

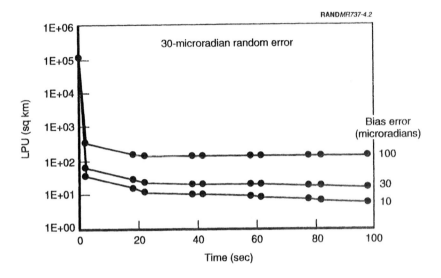

**Figure 4.2—Sensitivity of LPU ($\ell = 2$) to Bias Error
(Two Sensors; Suboptimal Filter)**

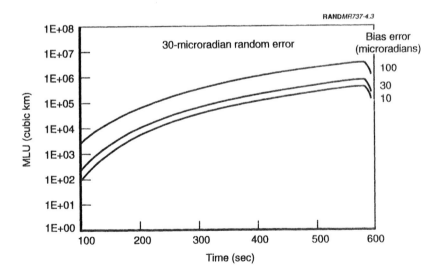

**Figure 4.3—Sensitivity of MLU ($\ell = 2$) to Bias Error
(Two Sensors; Suboptimal Filter)**

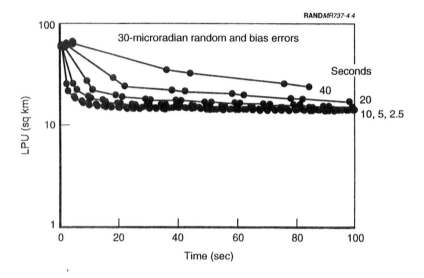

Figure 4.4—Sensitivity of LPU (ℓ = 2) to Revisit Time (Two Sensors; Suboptimal Filter)

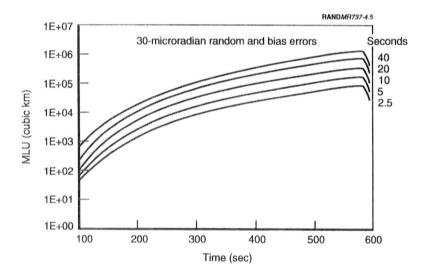

Figure 4.5—Sensitivity of MLU (ℓ = 2) to Revisit Time (Two Sensors; Suboptimal Filter)

statistics alone do not "beat down" the effects of bias. Although the insensitivity to revisit time is not apparent in the case of missile location uncertainty (Figure 4.5), the spread in MLU values with varying revisit time is reduced by roughly an order of magnitude relative to the unbiased case.

In summary, we show launch point and missile location uncertainties at apogee (327 sec into the trajectory) in Tables 4.1 and 4.2, respectively, for varying revisit times (i.e., numbers of measurements) and bias errors (treated suboptimally). Note that when the bias error is large, increasing the number of measurements can lead to larger uncertainties.

Although counterintuitive, such behavior might be expected from a suboptimal formulation, especially when the bias errors are large. In such a treatment, the filter applies gains—indeed, sometimes large gains—to the system by considering random errors alone [see Eq. (2.29)]. As a result, when the effects of bias are examined, they may be large because they are amplified by large gains. If the filter was aware that bias errors were present, it could adjust these gains accordingly.

These results underscore the importance of properly accounting for bias under such circumstances. Next we examine an alternative treatment that remedies inconsistencies by incorporating the bias terms into the filter directly.

Table 4.1

LPUs (ℓ = 2) for Two Sensors Processed in Stereo
(Suboptimal Filter[a], in km^2)

Revisit Rate (sec)	Number of Measurements	Bias Error(μrad)			
		0	10	30	100
40	5	10.9	12.4	23.7	150.7
20	10	4.8	6.2	17.2	142.7
10	20	2.4	3.8	15.1	143.4
5	40	1.2	2.7	14.2	145.7
2.5	80	0.6	2.1	13.9	147.9

[a]30-microradian random error.

Table 4.2

MLUs ($\ell = 2$) at Apogee for Two Sensors Processed in Stereo (Suboptimal Filter[a]; Equivalent Spherical Radii in km)

Revisit Rate (sec)	Number of Measurements	Bias Error(μrad)			
		0	10	30	100
40	5	30.5	31.0	33.8	45.7
20	10	22.4	23.4	28.5	49.7
10	20	15.8	17.1	22.2	38.0
5	40	11.2	13.0	17.6	29.5
2.5	80	8.0	10.1	14.1	23.5

[a]30-microradian random error.

OPTIMAL TREATMENT OF BIAS

We are interested in redesigning a filter for the linear system[5]

$$z - \tilde{z} = H(x - \tilde{x}) + v + b \,, \tag{4.1}$$

where x is a six-dimensional vector and the bias term is assumed constant but unknown. In our previous construction, bias was treated as inherently unobservable, with its effects in some sense modeled as an afterthought. In practice, however, it may be possible to learn about bias through the measurement sequence, which allows the filter to adjust the gain optimally, with consideration given to both random and bias errors.

To treat bias as an observable, we first incorporate it into the state vector.[6] For a single sensor, define

---

[5]For simplicity, the index i is suppressed in this section.

[6]See B. Friedland, 1969, p. 360.

$$\Xi \equiv \begin{pmatrix} x_1 \\ x_2 \\ x_3 \\ x_4 \\ x_5 \\ x_6 \\ x_7 \\ x_8 \end{pmatrix} = \begin{pmatrix} x_1 \\ x_2 \\ x_3 \\ x_4 \\ x_5 \\ x_6 \\ b_1 \\ b_2 \end{pmatrix} , \qquad (4.24)$$

where x_1 through x_6 are defined as before [see Eq. (3.16)], and

$$b_1 = \text{bias on measurement } z_1$$

$$b_2 = \text{bias on measurement } z_2 . \qquad (4.25)$$

Concurrent with this state vector redefinition, redefine the H-matrix as

$$H = \begin{pmatrix} \dfrac{\partial z_1}{\partial x_1} & \dfrac{\partial z_1}{\partial x_2} & \dfrac{\partial z_1}{\partial x_3} & \dfrac{\partial z_1}{\partial x_4} & \dfrac{\partial z_1}{\partial x_5} & \dfrac{\partial z_1}{\partial x_6} & 1 & 0 \\ \dfrac{\partial z_2}{\partial x_1} & \dfrac{\partial z_2}{\partial x_2} & \dfrac{\partial z_2}{\partial x_3} & \dfrac{\partial z_2}{\partial x_4} & \dfrac{\partial z_2}{\partial x_5} & \dfrac{\partial z_2}{\partial x_6} & 0 & 1 \end{pmatrix}_{x = \tilde{x}} . \qquad (4.26)$$

Since the bias enters the problem as an exactly linear term, there is no need to generate a bias estimate for the purpose of Taylor expansion [see Eq. (2.18)]. Thus, Eq. (4.1) is rewritten as

$$z - \tilde{z} = H(\Xi - \tilde{\Xi}) + v , \qquad (4.27)$$

so that the form of the filter without bias is regained.

As a result, the framework of Chapter Three may now be applied to the bias problem by simply augmenting the state vector and H-matrix with additional terms. If we wish to model time-dependent errors associated with the choice $((z_1 , z_2) = ((\alpha_v , \Omega)$, [see Eq. (4.18)], we may incorporate appropriate factors into H:

$$H = \begin{pmatrix} \dfrac{\partial z_1}{\partial x_1} & \dfrac{\partial z_1}{\partial x_2} & \dfrac{\partial z_1}{\partial x_3} & \dfrac{\partial z_1}{\partial x_4} & \dfrac{\partial z_1}{\partial x_5} & \dfrac{\partial z_1}{\partial x_6} & 1 & 0 \\ \dfrac{\partial z_2}{\partial x_1} & \dfrac{\partial z_2}{\partial x_2} & \dfrac{\partial z_2}{\partial x_3} & \dfrac{\partial z_2}{\partial x_4} & \dfrac{\partial z_2}{\partial x_5} & \dfrac{\partial z_2}{\partial x_6} & 0 & g \end{pmatrix}_{x=\bar{x}}. \tag{4.28}$$

Finally, accounting for the bias errors of multiple sensors requires additional dimensions. For the case of stereoscopic processing (with time-dependent errors in Ω), Eqs. (4.24) and (4.28) generalize as

$$\Xi \equiv \begin{pmatrix} x_1 \\ x_2 \\ x_3 \\ x_4 \\ x_5 \\ x_6 \\ b_{11} \\ b_{21} \\ b_{12} \\ b_{22} \end{pmatrix}, \tag{4.29}$$

where

b_{11} = bias on measurement z_1, sensor 1

b_{21} = bias on measurement z_2, sensor 1

b_{12} = bias on measurement z_1, sensor 2

b_{22} = bias on measurement z_2, sensor 2, $\tag{4.30}$

and

$$H = \begin{pmatrix} \dfrac{\partial z_1}{\partial x_1} & \dfrac{\partial z_1}{\partial x_2} & \dfrac{\partial z_1}{\partial x_3} & \dfrac{\partial z_1}{\partial x_4} & \dfrac{\partial z_1}{\partial x_5} & \dfrac{\partial z_1}{\partial x_6} & 1 & 0 & 1 & 0 \\ \dfrac{\partial z_2}{\partial x_1} & \dfrac{\partial z_2}{\partial x_2} & \dfrac{\partial z_2}{\partial x_3} & \dfrac{\partial z_2}{\partial x_4} & \dfrac{\partial z_2}{\partial x_5} & \dfrac{\partial z_2}{\partial x_6} & 0 & g_1 & 0 & g_2 \end{pmatrix}_{x=\bar{x}} \tag{4.31}$$

with the subscripts 1 and 2 on the time-dependent factors referring to sensors 1 and 2, respectively.

Notional Results

When the bias is less than or comparable to the random error, one would expect the optimal and suboptimal formulations to yield roughly similar results. As a comparison of Figures 4.6 and 4.7 with Figures 4.1 and 4.2 illustrates, this is indeed the case for launch point uncertainty in the example presented here.

More prominent differences emerge in the missile location uncertainties. As Figure 4.8 depicts, MLUs calculated optimally are smaller and less sensitive to the bias magnitude than corresponding MLUs calculated suboptimally. In particular, MLU values in the case of 100-microradian bias errors are substantially reduced from the values shown in Figure 4.3, and all curves appear to coalesce late in the trajectory.

On intuitive grounds, one would expect such behavior: Since uncertainties late in the trajectory are dominated by "velocity-like" quantities (heading, loft angle, etc.), and since velocity can be discerned from differences in missile position, one would expect the constant bias errors to cancel. As a result, uncertainties late in the trajectory

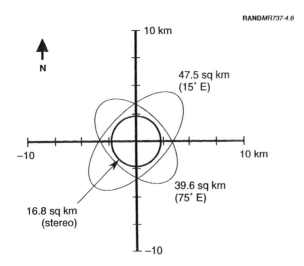

Figure 4.6—LPUs ($\ell = 2$) for Two Sensors with Random and Bias Errors (Optimal Filter)

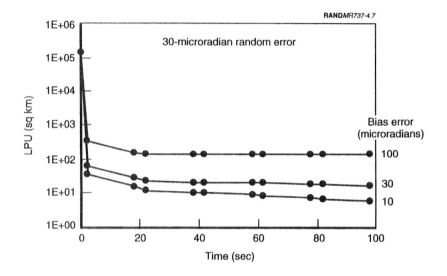

Figure 4.7—Sensitivity of LPU ($\ell = 2$) to Bias Error
(Two Sensors; Optimal Filter)

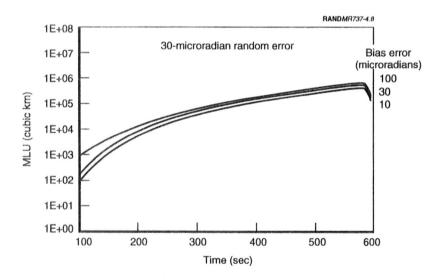

Figure 4.8—Sensitivity of MLU ($\ell = 2$) to Bias Error
(Two Sensors; Optimal Filter)

should be roughly independent of bias. This is not the case for two sensors using the suboptimal filter (Figure 4.3), but one can verify that with only one sensor this filter also yields curves that coalesce late in the trajectory.

As sensor revisit times are varied, the LPU calculated optimally (see Figure 4.9) again shows behavior similar to that calculated suboptimally, in the case when random and bias errors are comparable. On the other hand, MLUs demonstrate consistently smaller values in the optimal treatment (Figure 4.10), and also exhibit a greater sensitivity to the revisit time.

In summary, we show LPUs and MLUs at apogee in Tables 4.3 and 4.4, respectively, for varying revisit times and bias errors (treated optimally). In all cases, errors are reduced as the number of measurements increases. In addition, values are comparable to those obtained using the suboptimal approach when the bias is smaller than or comparable to the random error (see Tables 4.1 and 4.2).

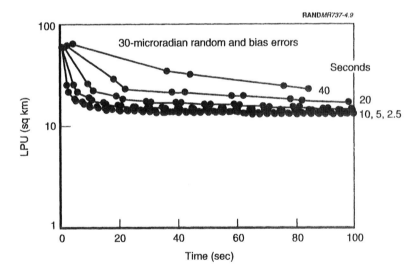

Figure 4.9—Sensitivity of LPU ($\ell = 2$) to Revisit Time
(Two Sensors; Optimal Filter)

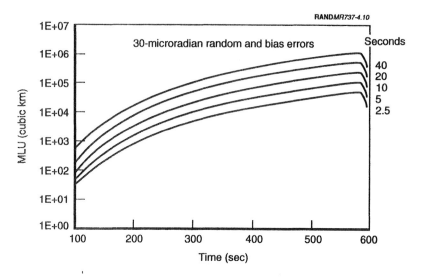

Figure 4.10—Sensitivity of MLU ($\ell = 2$) to Revisit Time
(Two Sensors; Optimal Filter)

Table 4.3

LPUs ($\ell = 2$) for Two Sensors Processed in Stereo
(Optimal Filter[a], in km^2)

Revisit Rate (sec)	Number of Measurements	Bias Error(μrad)			
		0	10	30	100
40	5	10.9	12.4	23.3	141.5
20	10	4.8	6.2	16.8	134.3
10	20	2.4	3.8	14.5	132.0
5	40	1.2	2.7	13.4	130.8
2.5	80	0.6	2.1	12.8	130.2

[a]30-microradian random error.

Table 4.4

MLUs ($\ell = 2$) at Apogee for Two Sensors Processed in Stereo
(Optimal Filter[a]; Equivalent Spherical Radii in km)

Revisit Rate (sec)	Number of Measurements	Bias Error(μrad)			
		0	10	30	100
40	5	30.5	30.9	32.8	35.4
20	10	22.4	23.3	25.6	27.7
10	20	15.8	17.0	19.3	21.1
5	40	11.2	12.8	15.0	16.7
2.5	80	8.0	9.8	11.8	13.5

[a]30-microradian random error.

Keep in mind, however, that the suboptimal treatment of bias can go awry even in instances when the random and bias errors are comparable. As mentioned previously, if the filter applies large gains during the estimation sequence, results obtained treating bias suboptimally may differ markedly from those obtained with an optimal formulation. In all cases, though, the former approach would *overestimate* the error.

In conclusion, treating bias as an observable quantity consistently yields smaller uncertainties than those obtained treating bias as an afterthought. Within the constraints of this limited analysis, then, our use of the terms "optimal" and "suboptimal" is clearly justified.

Chapter Five

CONCLUDING REMARKS

As theater missile defenses are fielded at the decade's end, satellite sensors will likely support vital TMD battle management functions. Waging "information warfare" will require increasingly sophisticated C^3I networks that are capable of piecing together the multifarious packets of information required to effect battlespace dominance. In this regard, timely transmission throughout the theater is central. But successful battle management requires more than connectivity alone: The *quality* of the information being transmitted is paramount. As we have shown, Kalman filtering of sensor measurements can in principle provide such information in the TMD operational environment. Thus, our primary focus has been on describing how to estimate the operational implications of this technique.

Table 5.1 summarizes some of the results from Chapters Three and Four for a notional missile launch against Israel. Using this table, it is possible to derive order-of-magnitude estimates for launch point uncertainties in a variety of situations not described explicitly in this report. For example, halving the TBM burn time to 50 sec would halve the number of measurements obtained, roughly equivalent to doubling the sensor revisit time. Including an opaque cloud deck at 10 km would eliminate measurements obtained during the first 42 sec of flight, again roughly equivalent to doubling the revisit time. Thus, a sensor with a 20-sec revisit time and 30-microradian random errors would yield an LPU of about 11 km^2 ($\ell = 2$, no bias) in both these situations.

In short, Table 5.1 is a useful guide to the *class* of numbers one would obtain in many situations of interest to TMD. Proper analysis of a

Table 5.1

LPUs (ℓ = 2) for Two Sensors Processed in Stereo[a]
(in km^2)

Revisit Rate (seconds)	Number of Measurements	Random Error (µrads)	Bias Error (µrad)			
			0	10	30	100
40	5	10	1.5	3.0	13.7	131.2
		30	10.9	12.4	23.3	141.5
		100	102.7	104.2	115.9	241.3
20	10	10	0.6	2.0	12.7	130.1
		30	4.8	6.2	16.8	134.3
		100	47.2	48.5	59.2	177.4
10	20	10	0.3	1.7	12.4	129.6
		30	2.4	3.8	14.5	132.0
		100	24.2	25.5	36.0	153.6
5	40	10	0.1	1.6	12.2	129.2
		30	1.2	2.7	13.4	130.8
		100	12.4	13.7	24.2	141.7
2.5	80	10	0.1	1.5	12.0	128.5
		30	0.6	2.1	12.8	130.2
		100	6.4	7.7	18.3	135.8

[a]Optimal filter.

different scenario requires doing the calculation correctly, as we have attempted here.

In addition, Table 5.1 underscores the point that *increasing the data collection rate will not reduce the launch point uncertainty significantly unless random errors dominate the measurement process*—statistics alone do not "beat down" the effects of bias.

In a similar vein, Table 5.2 summarizes the missile location uncertainty at apogee (327 sec after launch for the notional TBM), in terms of equivalent spherical radius. Because location is unique to a given missile trajectory, inferring approximate MLUs for missiles other than that used to generate this table is of limited utility. It is possible, however, to gain insights regarding MLUs for the same missile in an altered operational setting, such as one containing cloud cover.

Table 5.2

MLUs ($\ell = 2$) at Apogee for Two Sensors Processed in Stereo[a]
(Equivalent Spherical Radii in km)

Revisit Rate (sec)	Number of Measurements	Random Error (μrad)	Bias Error (μrad)			
			0	10	30	100
40	5	10	12.3	13.8	16.0	17.6
		30	30.5	30.9	32.8	35.4
		100	68.5	68.6	69.2	72.9
20	10	10	8.2	10.2	12.3	14.1
		30	22.4	23.3	25.6	27.7
		100	53.7	53.9	54.7	57.9
10	20	10	5.6	7.6	9.2	11.1
		30	15.8	17.0	19.3	21.1
		100	40.9	41.2	42.4	45.3
5	40	10	3.9	5.8	7.1	9.2
		30	11.2	12.8	15.0	16.7
		100	31.2	31.6	33.3	35.8
2.5	80	10	2.7	4.5	5.5	7.9
		30	8.0	9.8	11.8	13.5
		100	23.6	24.2	26.3	28.4

[a]Optimal filter.

With the exception of cases in which bias errors are significantly larger than random errors, these results (with bias treated optimally) would not change appreciably if the bias effects were calculated suboptimally. On the other hand, as discussed at the end of Chapter Four, there are cases when this approach yields markedly different error estimates from the optimal one, even with comparable random and bias errors. These cases notwithstanding, reasonable results in the suboptimal formulation may be expected so long as the bias is not large enough to drive the estimation sequence. If it is, an optimal formulation is preferred, if not required.[1]

---

[1]In any case, the suboptimal treatment yields an upper bound on the error. Depending upon the situation at hand, this may be sufficient information.

In conclusion, satellite sensors providing the type of information described in this report could, if harnessed in a theater of operations, enhance the capability of active defenses, passive defenses, and attack operations. As the above data demonstrate, this is especially true of sensors with short revisit times and small measurement errors, at least insofar as our notional trajectory analysis is concerned.

We have neglected a number of considerations, however, that could increase the uncertainty in our estimates. For example, our missile template has a fixed burn time that ignores uncertainty in the missile burnout velocity. What if the TBM did not burn its full 100 sec, but rather cut off its engines early? Since the acceleration near burnout is roughly 8 g's, each second of boost lost to early cutoff reduces the velocity by about 80 meters per second. With a velocity near burnout of roughly 3 km/sec, the change in TBM range resulting from early cutoff is therefore approximated by

$$\Delta r \approx \frac{2v\Delta v}{g} \approx 48\Delta t_{burnout} \text{ (km).} \qquad (5.1)$$

Thus, each second of nominal boost lost to early cutoff results in roughly 50 kilometers of range reduction. If the missile burned only 95 sec, its 1200-km trajectory would be reduced to roughly 950 km. It is therefore easy to see how this degree of freedom could easily dominate the error analysis.[2]

Other considerations might also be important, such as template errors (e.g., different TBM pitch angle profiles, nonplanar TBM motion), missile mistyping, effects from a rotating, nonspherical earth, process noise, or even missile maneuvering obscured by cloud cover.

As a final note, it is straightforward to adapt this methodology to different types of sensors. If interested in radars, for example, one would simply augment the matrices by an additional dimension to account for range measurements. Line-of-sight velocity measure-

---

[2]To provide an upper bound on the uncertainty in this case, one could join the two error ellipsoids generated by analyzing the full burn and early cutoff trajectories, respectively. For an alternative approach, see H. Holtz and L. R. Western, *Mathematical Assumptions in the MSTP Covariance Analysis*, El Segundo, California: The Aerospace Corporation, Report No. TOR-95(5411)-1, 1995.

ments could be incorporated similarly. In short, the approach is easily applied to a wide variety of sensors on spaceborne, airborne, sea-based, or ground-based platforms.

MISSILE TRAJECTORIES ON THE EARTH'S SURFACE

Ballistic missile trajectory data are most often specified in a generic, two-dimensional format (e.g., range and altitude versus time), without specific reference to the earth's surface. In this case, applying the methodology in the main text in a realistic operational setting requires transforming the data into a format appropriate for describing trajectories on the earth's surface. In what follows, we briefly describe how this may be accomplished.

Ballistic missile orbits trace out great circles on the surface of a non-rotating earth. For this reason, it is straightforward to employ spherical trigonometric relations to translate missile altitude and range (specified as a function of time on a curved earth) into a three-dimensional trajectory, for a given launch heading. Such a trajectory is illustrated in Figure A.1.

In Figure A.1, lower-case letters refer to angles subtended from the earth's center, whereas upper-case letters indicate interior angles of the spherical triangle abc formed by the intersection of great circles drawn on the earth's surface. More explicitly,

> a = complement of impact latitude
>
> b = instantaneous missile range/radius of earth (known)
>
> c = complement of launch latitude (known)
>
> A = launch heading angle (known)
>
> B = impact longitude—launch longitude.   (A.1)

Given a launch latitude, launch longitude, initial heading, and an instantaneous missile range, our goal is to solve for the corrresponding

RANDMR737-A.1

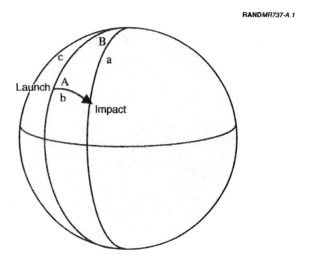

Figure A.1—Great Circles on the Earth's Surface

missile latitude and longitude at the same point in time along the
trajectory. (Without loss of generality, we refer to these as the "impact" latitude and longitude.) To accomplish this, we use the law of
sines,

$$\frac{\sin a}{\sin A} = \frac{\sin b}{\sin B},$$

(A.2)

as well as the law of cosines,[1]

$$\cos a = \cos b \cos c + \sin b \sin c \cos A.$$

(A.3)

With these equations, one can show that the latitude (l) and the
longitude (L) at impact are given by

$$l = \sin^{-1}\left(\cos b \cos c + \sin b \sin c \cos A\right)$$

(A.4)

---

[1]M. R. Spiegel, *Mathematical Handbook of Formulas and Tables,* New York: McGraw-
Hill, 1968, p. 19.

and

$$L = L_0 + \arg\left(\cos b - \cos c \sin l, \sin b \sin c \sin A\right), \qquad (A.5)$$

respectively, with L_0 the longitude at launch and $\arg(x, y)$ $= \tan^{-1}\left(y / x\right)$ if $x > 0$; $\arg(x, y) = \tan^{-1}(y / x) + \pi$ if $x < 0$; $\arg\left(x, y\right) = \pi / 2$ if $x = 0$, $y > 0$; $\arg(x, y) = -\pi / 2$ if $x = 0$, $y < 0$; and $\arg\left(x, y\right) = $ undefined if $x = 0$, $y = 0$.

Ballistic Missile Defense Organization, *Ballistic Missile Proliferation: An Emerging Threat*, Arlington, Virginia: System Planning Corporation, 1992.

Bryson, A. E., and Y.-C. Ho, *Applied Optimal Control*, New York: Hemisphere Publishing Corporation, 1975.

Carter, A., and D. N. Schwartz (eds.), *Ballistic Missile Defense*, Washington, D.C.: The Brookings Institution, 1984.

Congressional Budget Office, *The Future of Theater Missile Defense*, Washington, D.C.: U.S. Government Printing Office, June 1994.

Fetter, S., G. N. Lewis, and L. Gronlund, "Why Were Scud Casualties So Low?" *Nature*, 28 January 1993, pp. 293–296.

Friedland, B., "Treatment of Bias in Recursive Filtering," *IEEE Transactions on Automatic Control*, Vol. AC-14, No. 4, August 1969, pp. 359–367.

Gelb, A., ed., *Applied Optimal Estimation*, Reading, Massachusetts: The Analytic Sciences Corporation, 1974.

Goldstein, H., *Classical Mechanics*, Reading, Massachusetts: Addison-Wesley, 1980.

Holtz, H., and L. R. Western, *Mathematical Assumptions in the MSTP Covariance Analysis*, El Segundo, California: The Aerospace Corporation, Report No. TOR-95(5411)-1, 1995.

Kalman, R. E., "A New Approach to Linear Filtering and Prediction," *Trans. ASME*, Vol. 82D, 1960, p. 35.

Lennox, D., "Ballistic Missiles Hit New Heights," *Jane's Defence Weekly*, 30 April 1994, pp. 24–28.

Nolan, Janne E., *Trappings of Power: Ballistic Missiles in the Third World*, Washington, D.C.: The Brookings Institution, 1991.

Postol, T. A., "Lessons of the Gulf War Experience with Patriot," *International Security*, Vol. 16, No. 3, Winter 1991/92, pp. 119–171.

Reif, F., *Fundamentals of Statistical and Thermal Physics*, New York: McGraw-Hill, 1965.

Schwartz, D. N., *Past and Present: The Historical Legacy*, in A. Carter and D. N. Schwartz (eds.), *Ballistic Missile Defense*, Washington, D.C.: The Brookings Institution, 1984, pp. 330–349.

Secretary of Defense, *Conduct of the Persian Gulf War: Final Report to Congress*, Washington, D.C.: U.S. Government Printing Office, April 1992.

Spiegel, M. R., *Mathematical Handbook of Formulas and Tables*, New York: McGraw-Hill, 1968.

Stein, R. M., and T. A. Postol, "Correspondence: Patriot Experience in the Gulf War," *International Security*, Vol. 17, No. 1, Summer 1992, pp. 199–240.

Stein, R. M., "Patriot ATBM Experience in the Gulf War," *International Security*, Vol. 16, No. 3, Winter 1991/92, addendum.

Vaughan, D., J. Isaacson, J. Kvitky, and R. Mesic, *Evaluation of Operational Concepts for Countering Theater Ballistic Missiles*, Santa Monica, California: RAND, WP-108, 1994.

Waller, D. C., *The Commandos*, New York: Simon & Schuster, 1994.